The Musculoskeletal Manual

The Musculoskeletal Manual

JACOB S. BIRNBAUM, M.D.

Department of Family Practice
Kaiser-Permanente Medical Care Program
San Diego, California
and
Clinical Instructor
Department of Community Medicine
University of California
San Diego, California
and
Member, American College
of Sports Medicine

1982

ACADEMIC PRESS

DIVISION OF CONTINUING MEDICAL EDUCATION

A Subsidiary of Harcourt Brace Jovanovich, Publishers

New York   London
Paris   San Diego   San Francisco   São Paulo   Sydney   Tokyo   Toronto

ACADEMIC PRESS, INC.
111 Fifth Avenue, New York, New York 10003

United Kingdom Edition published by
ACADEMIC PRESS, INC. (LONDON) LTD.
24/28 Oval Road, London NW1 7DX

Library of Congress Cataloging in Publication Data

Birnbaum, Jacob S.
    The musculoskeletal manual.

    Bibliography: p.
    Includes index.
    1. Neuromuscular diseases--Handbooks, manuals, etc.
I. Title.
RC925.B5   616.7   82-3878
ISBN 0-12-788074-7   AACR2

PRINTED IN THE UNITED STATES OF AMERICA

82 83 84 85   9 8 7 6 5 4 3 2 1

Contents

Preface

There are many excellent texts and manuals available covering the diagnosis and management of severe trauma, such as fractures and dislocations, and equally fine books on the various rheumatologic, crystalline, neoplastic, and infectious conditions that affect the musculoskeletal system. However, primary care practitioners spend a fair portion of their time dealing with musculoskeletal complaints that do not fall into these categories—from tennis elbow to low back pain to children who walk pigeon-toed. Medical education, both undergraduate and graduate, has too often either ignored or lightly passed over these ubiquitous and sometimes disabling (although admittedly only rarely life-threatening) complaints. Furthermore, useful reference materials on common musculoskeletal complaints are hard to find. This manual is an attempt to remedy that. It is intended for family and general practitioners, internists and other specialists doing primary care, practitioners of sports medicine, family practice residents, medical students, physician's assistants, and nurse practitioners.

The text is divided into chapters primarily by body part, and the chapters are further subdivided by complaint. The method of evaluation of each complaint is then given, along with a discussion of the diagnosis, pathophysiology, and treatment of the conditions that can cause that complaint. Cross-references to other sections or chapters are often necessary, for obvious reasons.

Diagnostic techniques mentioned in the main body of the text are discussed more fully in Chapter 2 and therapeutic techniques in Chapter 3. If, for instance, antiinflammatory medication or steroid injection is advised for a certain condition, referral to Chapter 3 will give more information on the different antiinflammatory medications available and their relative advantages and disadvantages. It will also provide information on the contraindications and precautions that need to be taken into account when using steroid injections.

Lower extremity problems in runners are rapidly becoming more and more common and are discussed separately in Chapter 20. The evaluation and management of pediatric musculoskeletal problems is the topic of Chapter 19 and the final five chapters discuss muscle problems, degenerative arthritis, and the rheumatic and crystalline diseases and infections.

This book has been organized so as to present a step-by-step approach to the diagnosis and management of musculoskeletal conditions. It is of course impossible to anticipate or write about every possible variation in symptom, pathology, or management. As always, it remains the task of the health care provider to apply knowledge obtained from all sources to the proper care of the individual patient.

Although every effort has been made to ensure the accuracy and completeness of the drug information presented herein, neither the author nor the publisher can assume responsibility for errors or for information that became known after submission of the manuscript for publication. It remains the responsibility of the physician to check complete product information before prescribing or recommending any drug.

Sincere thanks must go to the many reviewers whose suggestions and criticisms were so instrumental in making this book what it is. Space limitations preclude me from mentioning them all, but I would be remiss in not naming those whose suggestions were especially helpful: Raymond M. Vance, MD, Mitchell L. Feingold, DPM, J. Michael Kelly, MD, and Judith A. Soulé, MD.

Thanks also to the many physicians of the Southern California Permanente Medical Group in San Diego, especially those in the Departments of Family Practice, Physical Medicine, and Orthopedics, from whom I have learned so much over the years.

The manuscript was excellently typed by Sally Birnbaum and Mike Friedman. Gregg Smith did the illustrations better than I could have dreamed and is also responsible for the fine cover design. And most of all, thanks to my very lovely wife Janine B. Zukotynski, MD, who was my number one reviewer and whose patience through the long hours of work on this project made it possible. Any inadequacies that remain are, of course, entirely my responsibility.

The Fundamental Precepts of Musculoskeletal Management

1. Always check neurovascular status distal to an injury.
2. Remember that bony tenderness can mean infection, neoplasm, or significant occult trauma and must be evaluated by X ray and whatever other techniques are indicated.
3. If more than one joint or area of the body is involved, consider that the patient may have a systemic disease.
4. Never forget that neurologic symptoms, and sometimes pain, can be due to vascular or CNS disease or to peripheral neuropathy as well as to radiculopathy.
5. Always keep in mind the possibility of referred pain.
6. Acute monoarthritis or oligoarthritis may mean a septic joint, which must be tapped to rule out infection; but never tap into a normal joint through infected skin.
7. Always ask about allergy before injecting or prescribing medication.
8. When the picture is not clear or does not progress as expected, X rays should be obtained.
9. Differentiate periarticular from intraarticular disease.
10. Check on tetanus immunization status with any open injury.
11. Until a definitive diagnosis can be made and specific therapy instituted, treat most acute musculoskeletal injury with cold, compression, elevation, and immobilization.
12. In children, always consider the possibilities of epiphyseal injury or that the child is battered.

1

Definitions and General Principles

A MUSCULOSKELETAL GLOSSARY

Structures

Apophysis: A growth region where a tendon inserts into bone.
Bone: The rigid, living structure that forms the framework upon which the rest of the musculoskeletal system is built and upon which it operates. It is made up of a matrix composed primarily of collagen fibers into which are impregnated various calcium salts. Within this structure are *osteocytes,* the living cells of bone, as well as a system of nutritional pathways. Bone is a living tissue,

1

constantly being remodeled in response to metabolic conditions and physical stresses; *osteoblasts* are cells that create new bone, whereas *osteoclasts* destroy bone. Covering all this, and firmly attached to it, is the *periosteum,* a sheath of dense connective tissue.

Bursa: A cystic structure that forms between surfaces that move over each other, probably to lubricate that movement.

Epiphyseal Plate: The plate of cartilage near the end of a long bone, where growth occurs in infants, children, and adolescents.

Joint: A connection between bones. The kind of joint usually thought of is a *diarthrodal joint,* which allows free movement in at least some directions. In this type of joint the opposing ends of bone are covered by *articular cartilage,* and surrounding the joint are ligaments called the *capsule.* Lining the inside of the joint capsule is the *synovial membrane,* which is epithelial tissue that produces the *synovial fluid* that lubricates the joint.

The other type of joint that is movable (though much less so than the diarthrodal) is the *amphiarthrodal joint,* which is composed of a disc of fibrocartilage between the bones, again covered by a ligamentous capsule.

Ligament: A structure made of dense regular connective tissue that connects bone to bone, around a joint or elsewhere.

Muscle: The contractile tissue that, under voluntary or involuntary nervous-system control, moves and/or stabilizes the various body parts.

Tendon: Made of tissue similar to that of a ligament, but connects *muscle* to bone.

Pathology

Apophysitis: Excessive traction on an apophysis leads to inflammation. Most common are Osgood-Schlatter's disease of the knee and Sever's disease of the heel.

Arthritis: Inflammation of one or more joints. Can be of autoimmune, viral, septic, crystalline, degenerative, or traumatic etiology.

Avascular Necrosis: A condition, most common in childhood, in which a loss of blood supply, whether idiopathic or due to trauma or other factors, leads to death of a small or large piece of bone. Most common are Legg-Perthe's disease of the hip and the various locations of Osteochondritis Dissecans.

Avulsion Fracture: A fracture in which only a small chip of bone is broken off, invariably at the insertion of either a ligament or (less commonly) a tendon. It can usually be treated the same as a third-degree sprain.

Bursitis: Inflammation of a bursa. There are two types: the superficial bursae (e.g., olecranon and prepatellar), except when infected or directly traumatized, usually present with painless swelling, while the deeper bursae (e.g., trochanteric and subacromial) present primarily with pain.

Compartment Syndrome: The increase of pressure in a fascial compartment of the arm or leg due to bleeding, infection, or tissue swelling. When acute it can lead to nerve compression and permanent neurologic sequelae if not treated emergently.

Connective Tissue Disease: One of various systemic diseases of unknown etiology. See Chapter 23.

Contusion (Bruise): A consequence of direct trauma to muscle or bone, in which there is some extravasation of blood around or in the muscle or under the periosteum.

Cramp: A painful, localized, transient spasm of muscle with a variety of possible etiologies. See Chapter 21.

Crystalline Disease: Gout and pseudogout; conditions in which the deposition of crystals in joints leads to pain and disability. See Chapter 24.

Degenerative Arthritis (or Degenerative Joint Disease): Degeneration of joint surfaces and other consequences that result from wear and tear or previous trauma. See Chapter 22.

Dislocation: An injury in which the normally articulating surfaces of a joint are displaced out of proper position. It is invariably associated with some degree of capsular ligament damage, and sometimes with a fracture as well.

Epiphyseal Injury: Slippage or crush of, or fracture through, the growth plate.

Fracture: A disruption in the continuity of bone.

Muscle Pull: See Strain.

Myositis Ossificans: A condition in which bone forms inside muscle tissue. See Chapter 21.

Nerve Compression Syndrome: The compression of a peripheral nerve, giving rise to pain and/or neurologic symptoms. The Carpal Tunnel Syndrome is the most common.

Osteochondritis Dissecans: A condition in which a small piece of articular cartilage and attached bone loses its blood supply, necroses and breaks off, and becomes a loose body in the joint. Most common in the knee and elbow.

Pathologic Fracture: A fracture through diseased bone, often bone involved with neoplastic disease. This possibility should be considered whenever the trauma reported by the patient seems to be too minor to produce the observed fracture.

Periostitis: Inflammation of the periosteum is caused by excessive traction there by ligaments or more commonly, tendons or muscle origins. (Some degree of periosteal inflammation accompanies many cases of tendonitis, especially when at muscle origin rather than insertion: lateral epicondylitis of the elbow and some forms of shin splints, for example).

Radiculitis: Inflammation of a nerve root; usually leads to localized as well as radiating pain.

Radiculopathy: Nerve root impingement and/or inflammation that has progressed to the point where neurologic symptoms or findings are present in the areas innervated by the involved root.

Sprain: An injury to a ligament or group of ligaments. First degree sprains involve only excessive stretching, while second degree sprains are partial tears and third degree sprains are complete tears of the ligaments.

Strain: This is a stretching injury to muscle. Degrees of injury are defined analogously to sprains, discussed above. Proper warm-up and stretching are essential for its prevention.

Stress Fracture: Bone is a dynamic tissue, constantly remodeling itself. When, due to excessive stress at a particular point or points, osteoclastic activity becomes more intense than osteoblastic activity, the bone is weakened. (The precise mechanism of this process is unknown.) At what point one stops calling it a stress injury and starts calling it a stress fracture is a matter of semantics; suffice it to say that a stress fracture (or severe stress injury) can, if not properly treated (mainly by desisting from the offending activity), progress in sudden and dramatic fashion to a complete through-and-through fracture. Patients with unusual or repeated stress fractures should undergo a workup to rule out underlying metabolic disease (BUN, creatinine, calcium, phosphorus, protein, acid phosphatase, alkaline phosphatase, and electrolytes).

Stress Injury: A milder form and antecedent of a stress fracture. See above.

Subluxation: An incomplete dislocation.

Tendonitis: Simple inflammation of a tendon.

ARTICULAR VERSUS PERIARTICULAR VERSUS REFERRED PAIN

The majority of patients presenting with musculoskeletal pain will refer to a particular joint or joints as the source of affliction. The most basic diagnostic decision that has to be made is whether the disability is articular in origin, i.e., within the joint itself, or rather actually arises from one or more of the various structures (tendons, ligaments, bursae, muscles) that surround the joint. Location of tenderness often makes the distinction simple. Another point is that intrinsic joint disease of whatever etiology usually (but not always) gives pain upon movement of the joint in all directions, while periarticular conditions will lead to discomfort upon movement only in certain planes. The differential diagnosis of articular disease is discussed briefly in the sections that follow, and more completely in Chapters 22–25.

Note that if a relatively pain-free range of motion is found and no significant tenderness can be elicited, one must consider the strong possibility of pain referred from other structures, musculoskeletal or otherwise.

MONOARTHRITIS: DIFFERENTIAL DIAGNOSIS

The single acutely and atraumatically inflamed joint must be considered to be infected until synovial fluid analysis proves otherwise. In other words, the joint must be tapped.*The technique for this procedure is discussed in the chapters on the individual joints; and synovial fluid analysis is discussed in Chapter 2. Never make the mistake of tapping into a normal joint through infected skin.

A septic joint is an emergency that must be treated aggressively if serious disability is to be avoided. See Chapter 25.

Another common cause of an acute monoarthritis is crystalline disease, more often gout than pseudogout. These conditions are discussed in Chapter 24.

Acute exacerbations of degenerative arthritis, often precipitated by overuse, can result in a similar picture. Rheumatoid arthritis or other rheumatic diseases can sometimes begin in one joint in a fairly acute manner. See Chapter 23. And, very rarely, monoarthritis may signify a primary neoplasm of the joint.

OLIGOARTHRITIS: DIFFERENTIAL DIAGNOSIS

The leading possibilities are gonococcal arthritis and the arthritis associated with various viral illnesses (especially hepatitis, mononucleosis, and rubella). Rheumatoid arthritis may begin this way, especially if small joints are involved fairly symmetrically. Related diseases and rheumatic fever should also be considered, the latter especially if the arthritis is migratory. See Chapter 23. An exacerbation of generalized degenerative arthritis related to overactivity is another potential cause.

POLYARTHRITIS: DIFFERENTIAL DIAGNOSIS

See the discussion at the beginning of Chapter 23.

RADICULAR PAIN

Aside from the specific evaluation discussed in Chapters 5 and 13, two possible etiologies deserve consideration:

---

*Exceptions might be an acutely inflamed small joint in a patient with a past history of gouty attacks, or an acutely swollen first MTP joint in a patient found to have an elevated uric acid level and without a nearby break in the skin.

1. *Preeruptive* (or rarely noneruptive) *herpes zoster* can give a severe, superficial burning type of pain.
2. A *diabetic mononeuropathy* can produce a radicular pattern of pain, occasionally in a previously undiagnosed diabetic. Rarely, several nerves can be involved.

MANAGEMENT OF ACUTE MUSCULOSKELETAL INJURY

1. Obviously, serious injury to other organs must be considered and dealt with first.
2. Check to be sure that there is no neurovascular compromise distal to the injury.
3. Fractures and dislocations should be immobilized pending definitive treatment. The reader is referred to any of the excellent textbooks or manuals of orthopedic or emergency medicine.
4. Minor musculoskeletal injury (bruises, strains, and sprains) are treated acutely with the ICCE regime:
Immobilization
Compression
Cold
Elevation
5. With open injuries, remember to check on the tetanus immunization status of the victim (see Table 1-I).

CHRONIC PAIN

This of course is a subject in itself, but since the majority of these patients complain of musculoskeletal pain, it is briefly mentioned here. Patients should only be placed into this category when an extensive evaluation has failed to reveal a specifically treatable cause for his or her misery. These patients usually have pain that is quite real, but of course psychological factors are often prominent, if not in the production of the pain, then certainly in the way the patient reacts to it. Thus a detailed psychological evaluation, especially looking for a masked depression or secondary gain, is indicated. Beyond this, a variety of modalities are utilized in the management of these patients. The transcutaneous nerve stimulator and hypnosis are currently the most widely used.

Table 1-I   Tetanus Prophylaxis

Clean wounds	
Immunized, booster within 5 years	None
Immunized, last booster more than 5 years previous	Booster[a]
Unimmunized	Course[b]
Tetanus-prone wounds (deep, extensive, contaminated, or more than six hours old)	
Immunized, booster within 1 year	None
Immunized, last booster between 1 and 10 years previous	Booster[a],
Immunized, last booster more than 10 years previous	Booster[a] 250–500 units TIG[c] in different arms[d]
Unimmunized	Course[b], 250–500 units TIG[c] in different arms[d]

[a] 0.5 cc tetanus toxoid.

[b] 0.5 cc tetanus toxoid, repeat in one month and then one year.

[c] Tetanus immune globulin, adult dose. Check on allergy first.

[d] Some authorities recommend a course of prophylactic antibiotics (penicillin or tetracycline) as well.

OTHER PRINCIPLES

1. Remember that bony tenderness can mean infection, neoplasm, or significant occult trauma, and must be evaluated by X ray and whatever other techniques are indicated.

2. If more than one joint or area of the body is involved, consider that the patient may have a systemic disease.

3. Never forget that neurologic symptoms and sometimes pain can be due to vascular or CNS disease or to peripheral neuropathy as well as to radiculopathy.

4. Always keep in mind the possibility of referred pain.

5. Always ask about allergy before injecting or prescribing medication.

6. When the picture is not clear or does not progress as expected, X rays should be obtained.

7. Check on tetanus immunization status with any open injury.

8. In children, always consider the possibilities of epiphyseal injury or a battered child.

2

Diagnosis

HISTORY

1. In patients who complain of musculoskeletal pain or swelling, ask about the *duration* and *consistency* of the symptom, as well as about *precipitating and relieving factors* (position, activity, mental stress, medications). Has there been any recent *overuse*?

2. Inquire as to whether there has been *any trauma or strain* to the involved part. If so, the specific mechanism of injury, if remembered, can be helpful in making a diagnosis. Also ask about whether there was a tearing or popping sensation at the moment of injury, how soon the part became swollen, and if there was any ecchymosis. Remember that injury in the past may not be recognized by the patient as the cause of the current affliction, and should be inquired about.

3. Especially when the neck or back is involved, ask if there is any *radiation of the pain, weakness, numbness, or paresthesia* distally. Conversely, neurologic complaints or diffuse pain in the extremities should lead to questioning about symptoms in the neck or back.

4. It is important to inquire about *symptoms in other joints or body parts;* a positive response certainly does not prove the presence of a systemic illness such as rheumatoid arthritis or polymyalgia rheumatica, but it must lead to consideration of such a possibility and possibly to further testing. Does the patient have a previous history of arthritis or gout?

5. Ask about *locking* or restricted range of motion of a joint, and if it is a weight-bearing joint, about *collapsing* (giving way) as well.

6. Find out about *previous episodes* of symptoms in the involved area, and what diagnoses were made and the effectiveness of any treatment given.

7. Other inquiries that should be made regarding specific complaints are mentioned in the corresponding sections in the text.

PHYSICAL EXAMINATION

This is discussed in the sections in the text on individual complaints, but a few general points will be made here.

1. Observe the functional status of the part in question: For example, whether the patient can walk, or how well he can use his hand.

2. Always try to localize *tenderness*. Bony tenderness should bring to mind the possibility of infection, tumor, or significant occult trauma, and the area should be X-rayed and possibly even a bone scan performed. The exception to this rule would be tenderness at commonly inflamed sites of attachment of tendons and ligaments (epicondylitis, for instance), but even in these locations a lack of expected improvement and timely resolution should lead to further evaluation.

3. *Swelling* of the joint space itself must be differentiated from the soft tissue swelling of acute injury and from effusion in superficial bursae. Joint swelling implies intraarticular disease or trauma, very possibly of a significant nature.

4. Any *redness* or *warmth* implies an acute inflammatory process.

5. Test and record the *range of motion* of any involved joints. Pain in all directions of motion usually means an articular process; periarticular conditions usually limit motion in only certain directions.

6. Always check *distal neurologic* (strength and sensation) and *vascular* (pulses, temperature, color and capillary filling) *status.*

7. Of course, a complete examination will not be needed on every patient, but it is much better to do too much than too little.

X RAYS

A few general guidelines for when X rays should be obtained as part of the management of musculoskeletal conditions are presented here. It must be noted, however, that this remains a topic in which great variation of opinion and practice still exists.

1. Take an X ray of any significant trauma.
2. Patients with typical presentations of common syndromes (e.g., epicondylitis, rotator cuff tendonitis, chondromalacia patellae) need not be X-rayed, but this decision should be reconsidered if the condition does not evolve and improve as expected.
3. Always take an X ray of children who complain of musculoskeletal pain, as the possibility of a catastrophic condition is significantly higher than in adults.
4. Bony tenderness (except at typical sites of insertional strain) or deep nonlocalizable pain should lead to radiographs being taken.
5. In patients above the age of 50 or so, degenerative musculoskeletal conditions become more common, but then so does the possibility of neoplastic disease. (A good percentage of cancers first present with the bony pain of metastasis.) Thus the threshold for obtaining X rays should lower with advancing age.
6. X rays of the low back and hips involve a significant radiation exposure to fairly radiation-sensitive tissues such as the gonads and kidneys, and so criteria are somewhat different; see the discussion in Chapters 11 and 14.
7. X ray the suspected source (neck or back) of distal radiculopathy.
8. X rays may sometimes have a therapeutic effect, by allaying a patient's expressed or unexpressed fear of serious occult disease (especially cancer); their use for this purpose in such a situation is certainly reasonable.
9. Always check to be sure that the patient is not pregnant or possibly pregnant before taking an X ray. This should heighten the threshold for ordering extremity X rays, but if still indicated they should be done, using a pelvic shield. X rays of the low back or hips should be deferred if at all possible.
10. If suspicion of neoplasm or infection is high and X rays are negative, order a bone scan (discussed below).

ARTHROGRAPHY

This is a technique wherein contrast material is injected into a joint (most commonly the knee) in which an internal derangement is suspected. Various intraarticular conditions invisible on a plain X ray (such as loose bodies or torn menisci) can be diagnosed by this method with a variable level of certainty. It is an invasive and fairly painful procedure, and probably should be limited to cases in which disability is great enough so that the patient would agree to surgery if indications for operation were found. As arthroscopy techniques improve, less arthrography is being done, but it still has its place in the diagnosis of certain conditions.

ARTHROSCOPY

As the technology of fiber optic arthroscopes improves and orthopedists gain more experience with this technique, it is becoming more commonly used in the knee and in some other joints as well. This procedure is considerably more accurate than arthrography (with some exceptions), but has the disadvantage of requiring anesthesia and a small incision, and of being more invasive. Perhaps its greatest benefit to date is in the avoidance of unnecessary arthrotomies. (Commonly patients are now giving consent to arthroscopy, with the understanding that the procedure will be expanded to an arthrotomy if indications are found.) Some therapeutic measures (such as the removal of small foreign bodies) can now be performed with this method as well.

MYELOGRAPHY

The diagnosis and preoperative evaluation of protruding intervertebral discs or spinal cord tumors often require this technique, in which a lumbar puncture is performed and contrast material (or sometimes air) is injected into the spinal canal. The patient is then tilted to allow the contrast medium to outline the suspected source of pathology. This procedure is necessary when significant or progressive neurologic symptoms are present or when the possibility of a cord neoplasm is being entertained, and is sometimes used also in the evaluation of severe and prolonged back pain. Another method sometimes used for the same purpose is *lumbar venography*. The development of CT scanning (discussed next) may soon greatly decrease the use of both of these techniques.

CT SCANNING

This technology, which has so revolutionized the diagnosis of intracranial pathology, is now being applied to various other body parts, including the spine. Accuracy in the diagnosis of disc herniations and intracanal masses is improving quickly, and this noninvasive method may soon replace myelography in a great many cases.

BONE SCANNING

In this technique a radioactive tracer (usually technetium-99 attached to one of various complex phosphate ions) is injected intravenously, and the whole skeleton or the part of interest is then imaged one to two hours later. Focal tracer accumulation is most often secondary to uptake in reactive bone formation, as in that associated with infection, fracture or neoplasm, but can also occur by various other mechanisms, such as inflammation of adjacent soft tissue with increased blood flow. This is an extremely useful method for detecting bony lesions of serious import long before changes are visible on X ray, and should be considered whenever the possibility of such a lesion is being considered and the X ray is normal.

EMG AND NCV

These techniques are often extremely helpful in differentiating the possible causes of muscular weakness, numbness and paresthesias.

Electromyography (EMG) consists of the insertion of thin electrode needles into the various muscles to be studied, and the observation of the motor unit potentials on an oscilloscope screen. Various abnormal potentials and patterns are seen with denervation of the muscle. The specific location of the pathology (root compression, compression of a specific peripheral nerve at a specific site, or a polyneuropathy) can then be surmised from the pattern of involvement of various muscles whose innervations are known. Intrinsic muscular disease or myesthenia gravis will give their own specific electromyographic patterns.

In measuring *nerve conduction velocity* (NCV) a stimulus is applied to a peripheral nerve at a specific site, and the time necessary for muscle action potentials to be picked up distally is measured. These are then compared to normal values for the age and sex of the patient; a delay implies peripheral nerve compression somewhere in between. This technique is especially useful to confirm a clinical impression of the location of compression before surgery is performed. (Some polyneuropathies slow the conduction velocity while others do not, but a comparison with other nerves will help differentiate these from compression neuropathies.)

Table 2-I   Synovial Fluid Cell Counts

	Normal	Noninflammatory	Inflammatory	Septic
Appearance	Clear	Cloudy	Turbid	Very turbid or purulent
White cell count (per cc) (some overlap is possible)	<500	500–5000	5,000–50,000	>50,000
White cell differential	Mostly monos		Mostly polys	

SYNOVIAL FLUID ANALYSIS

This must be performed whenever the possibility of infection is considered. (Techniques for specific joints will be found in the corresponding chapters. Be careful never to tap into a normal joint through infected skin.) It is also the only way to confirm definitely the diagnosis of gout or pseudogout and in addition is useful in differentiating essentially noninflammatory joint diseases (such as degenerative arthritis or an effusion due to overuse) from inflammatory conditions (such as rheumatoid arthritis). Several things must be checked for in the fluid obtained.

Appearance: Normal joint fluid is straw-colored and clear. Blood indicates either significant intraarticular trauma or a bleed into a joint due to hematologic disease (assuming of course an atraumatic tap). Slightly cloudy fluid can be seen with degenerative arthritis or overuse; turbid fluid indicates an inflammatory condition; very turbid or purulent fluid is seen with infection.
White Cell Count and Differential: See Table 2.I.
Crystals: The presence of negatively birefringent crystals means gout (uric acid); positively birefringent, pseudogout (calcium pyrophosphate).
Gram Stains and Cultures: Gram stains and cultures should always be obtained. Gram stains can be falsely negative, and the cell counts seen with inflammatory and septic arthritis can overlap; if any doubt exists the patient should be treated as if he had a septic arthritis (Chapter 25) until cultures prove otherwise.

BLOOD TESTS

Sedimentation Rate: This is an excellent screening test for systemic disease in a patient with multiarea involvement. It will be increased in rheumatic fever, rheumatoid arthritis, the rheumatoid variants, SLE, and polymyalgia rheumatica, as well as with acute infection, and can also be used to follow the level of activity of these conditions. It is not increased in crystalline disease (except during severe acute attacks) or degenerative arthritis.

The ESR increases with normal aging; normal values vary with the specific technique used and the laboratory performing the test. Note that this is very much a nonspecific marker, and a persistent increase may well have a significant etiology other than those discussed above.

Complete Blood Count: Whereas peripheral leukocytosis can certainly be seen with acute joint infection, a normal white count (like a normal ESR) by no means rules it out.

A finding of anemia can be a clue to the presence of systemic illness such as SLE.

Rheumatoid Factor: This is a test for various antibodies to IgG. It may be negative in a fair number of patients with rheumatoid arthritis, especially when early or mild, and also is usually negative in the rheumatoid variants. Note that it may be positive in a fair percentage of patients with SLE, and in some other conditions (such as syphilis and SBE) as well. Also keep in mind that a positive rheumatoid factor in low titer can be present normally in older patients.

Antinuclear Antibodies: These are antibodies to various nuclear constituents (DNA, nucleoproteins, etc.) almost invariably found in patients with SLE. The test can also be positive in patients with scleroderma or Sjögren's syndrome, with various drug-induced lupus-like syndromes, and in about a quarter of patients with rheumatoid arthritis. The LE cell preparation is not as sensitive or as specific as the ANA test.

Uric Acid: The higher the uric acid level in the blood, the more likely the patient is to have one or more attacks of gout. The demonstration of a serum level above defined normal limits, however, does not unequivocally prove that a patient's symptoms are due to gout. See Chapter 24.

Serum Calcium, Phosphorus, Alkaline Phosphatase, Acid Phosphatase, and *Serum Protein Electrophoresis* are useful tests in detecting and discovering the cause of bone disease or lesions. Remember that normal alkaline phosphatase levels in children and adolescents are much higher than those in adults.

HLA-B27 Antigen: This marker is found in 90% of patients with ankylosing spondylitis, and in about 70% of those with Reiter's syndrome, psoriatic spondylitis or enteropathic arthritis. It can be present in normals as well.

3

Therapeutics

REST AND IMMOBILIZATION

Rest of an injured or inflamed part is a critical measure, necessary to allow healing to take place. This may mean a variety of different things, however.

1. Limitation of use of inflamed tissues is quite important. But the precise motions to be avoided must be explained to the patient, and often he or she must be given tips on how to perform daily tasks without further injuring the painful structure. For example, in lateral epicondylitis (tennis elbow), if the patient takes your admonishment "rest the elbow" to mean limiting activities requiring flexion and extension of that joint, he or she will not be avoiding stress to the inflamed tendon insertion; it is resisted extension of the wrist and supination of the forearm that must be avoided.

2. An area may be injured acutely and require certain measures not directly

related to the mechanism of injury (bedrest for a herniated nucleus pulposus, for example).

3. In the patient with neck or back pain in which emotional factors (even if nothing more than anxiety about the injury itself or its consequences) play a part, the relief of stress (resulting from time away from work, for example, or from allayment of financial worries) can give the best rest that the neck or back has had in a long time.

Immobilization of an injured part is necessary in severe injury, and often helpful in milder trauma as well. But the longer a part is immobilized, the more likely it is that the soft tissues will become stiff, the muscles around it lose their tone, and the immobilized joint lose proprioceptive sensation. These changes are often not easily reversible, and sometimes they are permanent. If an injury is known to require a certain duration of immobilization to heal properly, fine; but by all means never immobilize a joint that does not need to be immobilized (the PIP joint in the case of mallet fingers, for example), and keep the duration of splinting or casting to a minimum. Clear guidelines should be given patients about the rapid tapering of the use of mainly symptomatic immobilization (cervical collars often fall into this category, for instance); and when immobilization is discontinued, often a period of mobilization and/or strengthening exercises will be needed.

Prevention of further injury to a vulnerable structure is another measure that falls into this category of musculoskeletal therapeutics. For example, patients with recurrent ankle sprains should be advised to avoid irregular or banked surfaces, and can be taught how to tape their ankles for support; patients with a torn meniscus should be told not to take part in body-contact sports or in activities requiring sudden direction changes.

EXERCISES

There are three types of exercises to be considered.

1. *Aerobic exercise.* This is a general term to describe exercise intended to increase the patient's cardiovascular fitness. It may range from a gradual progressive program of daily walking to cycling, swimming, running, or other sports. In addition to its cardiac and respiratory benefits, it can have quite salutary psychological effects as well, often helping to relieve anxiety and depression and promoting a sense of general well-being. For these reasons it is to be recommended to all patients; in regard to musculoskeletal pathology, though, it is often specifically helpful in patients with chronic neck or back pain. Obviously, aerobic activities must be

chosen that do not exacerbate the patient's primary complaint and are consistent with his or her age and general physical condition.

2. *Flexibility exercises.* These are designed to restore a normal range of motion to body parts in which disuse has led to stiffness, and to prevent muscle pulls and tendon injuries that can result from stresses to a muscle group that is too tight. Stretching exercise programs for specific injuries and activities will be found throughout the book. Flexibility increases as a muscle is used; this is the reason that warming up is so important in the prevention of injury.

3. *Strengthening exercises.* These have a variety of objectives: to restore strength to muscles weakened by disuse secondary to injury or therapeutic immobilization; to help restore proper biomechanics in situations where lack of muscular tone contributes to malalignment and to stresses or injuries to other structures (as in quadriceps strengthening exercises for patellofemoral problems, e.g.); to prevent sprains to the ligaments that surround joints by strengthening muscular support to the joints; and to achieve a proper balance of strength between opposing muscle groups. Again, methods for specific exercises are scattered through the volume.

HEAT AND COLD

These treatment modalities are probably effective primarily by exerting a counterirritant effect. In general, patients should use whichever of the two gives them the most relief. The application of cold does lead to vasoconstriction and therefore is recommended in acute injury and following steroid injection; heat may often be more beneficial in muscular pain and may help in muscle relaxation. The immediate contrast between cold and heat application to a single area is a technique that sometimes gives more relief than either modality used alone. Topical counterirritant medicaments do often provide some symptomatic relief, and some of those available are listed in Table 3-I. Also see the discussion of physical therapy modalities in the following section.

Heat

Wet heat seems to have a somewhat better effect than dry heat. Local wet heat may be obtained in various ways (see Patient Handout 3-1). In any case, temperatures should never be high enough to cause significant pain or erythema of the skin. The patient should generally use local heat for no more than 20 to 30 minutes at a time, and no more than four times a day or so.

Table 3-I   Examples of Counterirritant Topical Medications[a]

Product	Ingredients
Aspercreme (Thompson)	Triethanolamine salicylate
Ben-Gay (Leeming)	Methyl salicylate, menthol
Deep-Down (J. B. Williams)	Methyl salicylate, menthol, methyl nicotinate, camphor
Heet (Whitehall)	Methyl salicylate, camphor, oleoresin capsicum, alcohol
Heet Spray (Whitehall)	Methyl salicylate, camphor, menthol, methyl nicotinate
Icy Hot (Searle)	Methyl salicylate, menthol
Infrarub (Whitehall)	Histamine dihydrochloride, oleoresin capsicum
Myoflex (Warren-Teed)	Triethanolamine salicylate
Panalgesic (Polythress)	Methyl salicylate, aspirin, menthol, camphor

[a] These preparations are available over-the-counter and are widely used. This chart is for the reader's convenience only, and the listing is by no means complete. Neither recommendation of the products listed nor criticism of those not listed is implied. Always check complete product information before prescribing or recommending any drug.

Patients with hand arthritis often find that dipping their hands in molten paraffin (temperature of approximately 120°F.) gives a lot of relief. The paraffin of course then adheres to the skin and provides a prolonged heating effect. Set-ups are available for home use; the use of a thermometer to monitor temperature is mandatory.

Local application of heat can be hazardous in patients with decreased sensation or circulatory impairment, and should not be prescribed in these patients except under supervision. Heat is also contraindicated over areas of malignancy, and is inadvisable in acute injury (cold being the standard thermotherapy in the latter situation).

Taking a warm bath at the start or end of the day often is helpful in patients with diffuse musculoskeletal pain and/or stiffness.

Cold

If an ice bag is not available, have the patient place some ice cubes into a plastic bag and then wrap a thin towel around it. In general, cold should not be used for more than 15 to 20 minutes at a session, four times a day or so. (In acute injuries such as sprains, however, ice packs can be applied as often as every hour or two for the first 24 to 72 hours.) In muscular pain such as in the neck or back, having a spouse or acquaintance massage the area with the ice pack may be more helpful than just placing it there. Immediate contrast with wet heat often feels good. Ice should not be placed directly on the skin.

An area may also be cooled by spraying it with ethyl chloride or flurimethane. This is very useful in the office before injection or aspiration, and also when used by a physical therapist prior to massage, stretching or other manipulation.

The use of cold is contraindicated over areas of decreased sensation or vascular compromise, in patients with vasculitis and cold-induced urticaria, and over open wounds.

PHYSICAL THERAPY

Physical therapists can be extremely helpful in the management of patients with musculoskeletal problems, especially those with back or neck pain and those being rehabilitated from injury or surgery. They can teach exercises appropriate for the patient's condition, instruct the patient on how to avoid chronic or recurrent injury to the involved area, and advise the patient on the use of other modalities such as heat and cold. They often provide significant psychological support to patients with chronic pain and disability, and give the patient whose acute problem is slow to resolve the feeling that something is being done.

A variety of additional modalities is also employed, with great variation depending on the particular therapist's training, experience and biases. They include:

Infrared. The use of radiated heat probably has no advantage over directly applied heat, and has some disadvantages (mainly with regard to cost and safety).

Diathermy. This is a method of providing "deep" heat by the use of induced electrical currents. (Directly applied heat tends to be superficial because of the insulating effect of the high water content of the dermis.) Whether whatever extra benefit occurs is worth the high cost is problematic.

Ultrasound. Mechanical oscillating waves of high frequency are produced and transmitted to the tissues. Its salutary effect is most likely due to the production of heat; a direct beneficial effect from the vibratory forces is conceivable but as yet unproven. Its comparative efficacy also remains unproven, but there are patients who swear by it.

Electrotherapy. The direct application of small currents to the body has been reported to help in acute injury. The *Transcutaneous Nerve Stimulator*

(TENS) has been effective in a fair percentage of previously unresponsive chronic pain patients. Its high cost has been a drawback, but recently much cheaper, disposable units have become available and are being promoted for use in the pain of acute injury as well.

Massage. There is no question that massage by a trained individual aids in pain relief and muscular relaxation. Though much is known about technique, there is little understanding of the mechanisms by which it works.

Stretching. This can be done by the patient (after proper instruction), manually by the therapist, or with the aid of traction set-ups. It is often quite helpful.

Manipulation. This potentially extremely useful modality has been out of favor with the medical profession because of its intimate historical association with chiropractic, but there are signs that it is becoming more widely accepted. It seems clear that there would be certain situations where the application of proper physical stresses to a structure that has a disability of movement will make more sense than the administration of chemicals to the body that surrounds the structure.

ANTIINFLAMMATORY MEDICATIONS

Aside from rest and immobilization, this is probably the most-used treatment modality in soft-tissue pathology. These drugs have a significant analgesic effect, but more importantly they get at the source of pain in that they decrease the inflammatory response (by affecting prostaglandin metabolism). They are thus sometimes superior to purely analgesic medications, and have the added advantage of being nonaddicting. They have the disadvantage of being fairly expensive (except for the first-line drug, aspirin) and of having some common side effects, most commonly gastrointestinal.

Contraindications and Precautions

1. These drugs should in general not be used in pregnant patients or when the possibility of pregnancy exists, and should also be avoided in nursing mothers.
2. There is cross-allergy between aspirin and some of the other nonsteroidal antiinflammatories (NSAIs); a patient allergic to any one of them (including aspirin) should not be given any drug in the group without careful consideration.
3. Some of these drugs should be used with caution (if at all) in patients with blood clotting problems and those taking anticoagulant medications.

Table 3-II   Examples of Aspirin Preparations[a]

Product	Ingredients	Usual adult dosage
PLAIN ASPIRIN		
Aspirin (EMPIRIN, BAYER, various other brands)	ASA 325 mg	ii QID prn
ENTERIC-COATED ASPIRIN		
ECOTRIN (Menley & James)	ASA 325 mg	ii QID prn
ASPIRIN WITH ANTACID		
ALKA-SELTZER (Miles)	ASA 324 mg, citric acid 1000 mg, sodium bicarbonate 1904 mg	ii QID prn
ARTHRITIS PAIN FORMULA (Whitehall)	ASA 486 mg, $Mg(OH)_3$ 60 mg, $Al(OH)_3$ 20 mg	i q3h prn
ARTHRITIS STRENGTH BUFFERIN (Bristol-Myers)	ASA 486 mg, $MgCO_3$ 146 mg, AlGlycinate 73 mg	ii QID prn
ASCRIPTIN (Rorer)	ASA 325 mg, Mg/Al hydroxide 150 mg	ii–iii QID prn
ASCRIPTIN A/D (Rorer)	ASA 325 mg, Mg/Al hydroxide 300 mg	ii–iii QID prn
BUFFERIN (Bristol-Myers)	ASA 324 mg, $MgCO_3$ 97 mg, AlGlycinate 49 mg	ii q4h prn
CAMA-INLAY TABS (Dorsey)	ASA 600 mg, $Al(OH)_3$ 150 mg $Mg(OH)_2$ 150 mg	i q4h prn
TIME-RELEASE ASPIRIN		
BAYER TIME RELEASE (Glenbrook)	ASA 650 mg	ii q8h prn

ASPIRIN WITH OTHER INGREDIENTS

Many such preparations are available; patients should be advised to use them only if the therapeutic effect of the added ingredients (caffeine, antihistamines, acetamenophen, etc.) is desired.

[a] These preparations are available over-the-counter. This chart is for the reader's convenience only, and the listing is by no means complete. Neither recommendation of the specific products listed nor criticism of those not listed is implied. Always check complete product information before prescribing or recommending any drug.

4. All of these drugs can lead to gastrointestinal side effects, including frank ulceration. They should not be used in patients with active GI disease, and alternatives should be carefully considered in patients prone to such pathology.
5. This listing is by no means complete; always check the package insert before prescribing any of these drugs.

Choosing an Antiinflammatory Drug

Aspirin is by far the cheapest medicine in this group. (Note that while acetaminophen is just as good an analgesic and antipyretic, it does not have antiinflammatory action.) Aspirin does, however, have some disadvantages:

1. To maintain antiinflammatory blood levels, it must be taken at least four times a day (though there are some more expensive aspirin preparations that claim to be effective on a q8h dosage schedule).
2. It has more gastrointestinal side effects than some (but certainly not all) of the other NSAIs. This may be ameliorated to some extent by using enteric-coated aspirin or preparations which have added antacid. See Table 3-II. Other salicylate molecules are asserted to have significantly less gastrointestinal side effects as well: Some of these are listed in Table 3-III.

Table 3-III   Examples of Nonaspirin Salicylates[a]

Product	Ingredients[b]	Recommended adult dosage
Arthropan liquid (Purdue Frederick)	Choline salicylate	i tsp q4h prn
Disalcid (Riker)	Salicylsalicylic acid	ii TID or iii BID
Magan (Adria)	Mg salicylate	6 tabs per day in divided doses
Mobidin (Ascher)	Mg salicylate	i–ii TID to QID
Trilisate (Purdue Frederick)	Choline salicylate, Mg salicylate	i to iii BID (of 500 mg size)

[a] This chart is for the reader's convenience only, and the listing is by no means complete. Neither recommendation of the specific products listed nor criticism of those not listed is implied. Always check complete product information before prescribing or recommending any drug.

[b] The choline salicylate and salicylsalicylic acid preparations are reputed to have a lower incidence of gastrointestinal side effects.

3. Because it is a widely used medication available without a prescription, some patients will not believe it is going to work as well as a more exotic (and expensive) prescription. Many patients can be educated out of this bias, but some cannot. This may be unfortunate, but that is the way it is.
4. Note that the contraindications to aspirin (including allergy and pregnancy) can also be contraindications to other nonsteroidal antiinflammatories.
5. See Table 3-IV for a comparison of some of the advantages and disadvantages of the various nonsalicylate NSAIs available. Some of these drugs are not approved for use in children under age 12. Although they do not profess to be better than aspirin, there is no question that some patients respond idiosyncratically better to some antiinflammatory medications than to others; in a patient who is not benefitting sufficiently, it is reasonable to try the available medications one at a time until an acceptable level of relief is obtained. Several new drugs in this class are being introduced and marketed as analgesics.
6. *Steroids,* when administered orally, can have a variety of side effects, especially when administered for prolonged periods. Aside from occasional use in a rapidly tapering manner in patients with acutely herniated discs, they are not recommended.

STEROID INJECTION

Steroid injection (see Table 3-V) is often an extremely helpful measure; it deposits potent antiinflammatory medication at the site of pathology, and systemic side effects are usually insignificant. It is not without its deleterious effects and risks, however. The points that are made in the following sections must be considered and observed religiously if problems (sometimes quite serious) are to be avoided.

Precautions before Injection

1. Always obtain *informed consent.* The patient must be told of your judgment of the chance of a good result from injection and of alternative treatments available. But even more important, he or she must be aware of the risks of, and possible deleterious effects from, the procedure (listed below).
2. Be sure the patient is *not allergic* to anything you will be injecting. The local anesthetic will much more often be a problem than the steroid.
3. *Resuscitation* equipment and expertise must be rapidly available.

Table 3-IV   Examples of Nonsteroidal Antiinflammatories[a]

Drug	Usual adult dosage	Comment
Indomethicin (INDOCIN; Merck, Sharp and Dohme) 25 mg, 50 mg	Usually 25 mg TID may use 50 mg TID in first few days of acute pain in long term use BID or possibly even HS dosing may be sufficient	Seems to be more potent than some of the others, and useful for acute problems High incidence of GI upset, and fairly high incidence of CNS problems, especially headache
Sulindac (CLINORIL; Merck, Sharp and Dohme) 150 mg, 200 mg	150 mg BID or 200 mg BID	BID dose convenient, less GI upset than indomethicin (and possibly aspirin) Seems fairly potent
Naproxen (NAPROSYN; Syntex) 250 mg, 375 mg	250 mg BID or 375 mg BID	BID dose convenient Approved as analgesic by FDA
Ibuprofen (MOTRIN; Upjohn) 300 mg, 400 mg, 600 mg	300 to 600 mg TID or QID	Approved as analgesic by FDA Seems to work fairly well in more chronic problems Probably less GI effects than ASA
Fenoprofen (NALFON; Dista) 200 mg, 300 mg, 600 mg	300 or 600 mg QID 200 mg q4–6h for pain	Approved as analgesic by FDA
Tolmetin (TOLECTIN; McNeil) 200 mg, 400 mg	400 mg TID or QID	
Meclofenamate (MECLOMEN; Parke-Davis) 50 mg, 100 mg	50 or 100 mg TID or QID	New drug Reported to have a high incidence of diarrhea

Nonsteroidal antiinflammatories marketed as analgesics

Naproxen sodium (ANAPROX; Syntex) 275 mg	Up to 5 a day in divided doses	Do not use with naprosyn
Zomepirac sodium (ZOMAX; McNeil) 100 mg	½ to 1 tablet q4–6h	New

Phenylbutazone group

Phenylbutazone (BUTAZOLIDIN, BUTAZOLIDIN ALKA; Geigy; AZOLID,	Up to 600 mg per day in divided doses	Real though small risk of aplastic anemia; though potent and cheap, use not recommended. May be safe if limited to less than a week's use

(Continued)

Table 3-IV (*Continued*)

Drug	Usual adult dosage	Comment
AZOLID-A, USV; available generically) 100 mg With or without antacid		
Oxyphenbutazone (OXALID, USV; TANDEARIL, Geigy) 100 mg	Same as for phenylbutazone	Same as for phenylbutazone
Colchicine		
Colchicine 0.6 mg	See package insert	Classically used in acute gout Recently has been used in discogenic low back pain Use with caution

[a] This chart is for the reader's convenience only, and the listing may not be complete. Neither recommendation of the specific products listed nor criticism of those not listed is implied. Always check complete product information before prescribing any drug. Most of these drugs are best taken with food or antacid.

Table 3-V   Examples of Injectable Steroids[a]

ARISTOSPAN	(Triamcinolone hexacetonide; Lederle) 5 mg/ml or 20 mg/ml
CELESTONE SOLUSPAN	(Betamethasone[b] sodium phosphate, betamethasone[b] acetate; Schering) 6 mg/ml
DECADRON	(Dexamethasone[b] sodium phosphate; Merck, Sharp, & Dohme) 4 mg/ml or 24 mg/ml
DEPO-MEDROL	(Methylprednisolone acetate; Upjohn) 20 mg/ml or 40 mg/ml or 80 mg/ml
HEXADROL	(Dexamethasone[b] sodium phosphate; Organon) 4 mg/ml or 10 mg/ml
KENALOG	(Triamcinolone acetonide; Squibb) 10 mg/ml or 40 mg/ml

[a] This chart is for the reader's convenience only, and the listing may not be complete. Neither recommendation of the specific products listed nor criticism of those not listed is implied. Always check complete product information before injecting any drug.

[b] Dexamethasone and betamethasone are considered several times more potent than triamcinolone and methylprednisolone.

4. Check the package insert for other precautions and contraindications relative to the drug you are using.

Risks of Injection

(All of these are very rare, but you and the patient must be aware of them.)
1. Introduction of *infection*.
2. *Needle trauma* to a nerve or vascular structure. (Know where you are injecting, and always aspirate before injecting.)
3. *Allergic* reaction.

Possible Deleterious Effects of Injection

1. Steroid injections near tendons can weaken them, especially if repeated often. Use of the tendon can then lead to rupture.
2. Intraarticular instillation of steroid can accelerate cartilage deterioration and thus degenerative arthritis, again especially if repeated often. Short-term benefit may to some extent be at the price of long-term detriment.
3. The injection of large amounts of steroid can on rare occasions lead to systemic side effects.

Technique of Injection

1. *Clean* the area. If a joint is to be entered, complete sterile technique must be used throughout.
2. A topical spray of *ethyl chloride* immediately before injection is quite helpful.
3. Unless deep injection into a large joint such as the knee is being done, *mix* the steroid with local anesthetic (without added epinephrine) in the same syringe. The local will not only make the shot hurt less, but the relief noted by the patient after injection will tell you that the medication is in the right place.
4. *Never* inject under pressure; most commonly this means that the needle is in a tendon, and significant degeneration of that structure can result. And do not even inject *around* a major weight-bearing tendon such as the Achilles or patellar.
5. *Never* inject where there is even the remotest possibility of infection. If tapping a joint, do not inject steroid if the synovial fluid is anything but perfectly clear.

Instructions to the Patient after Injection

1. The patient should place ice packs on the area for 15 to 30 minutes four times a day for two days. This helps a great deal to prevent the flare of pain which often occurs soon after injection (possibly due to the crystallization of some of the steroid suspension).
2. If injection was near a tendon, sudden ballistic stress on that tendon (as in throwing a ball or swinging a racquet) should be avoided for several weeks.
3. The shot should not be considered a cure-all; the painful structure should still be rested, and other measures such as thermotherapy and oral medication should be used as appropriate.

PAIN MEDICATIONS

When possible, acetaminophen or aspirin or one of the other nonsteroidal antiinflammatory medications should be used. However, there will obviously be cases in which the analgesic effect of these drugs will be insufficient. In these situations, or when antiinflammatory drugs are contraindicated and the pain is too severe to be controlled by acetaminophen, one of a variety of analgesic medications can be prescribed. These more potent analgesics are more likely to result in various side effects, and also can lead to dependence if overused. But keep in mind that the relief of pain is quite important in breaking the pain → muscle spasm → pain cycle which so often prolongs soft-tissue distress, and so should be considered as a measure to help bring about resolution of the episode, as well as one to provide temporary relief from discomfort.

MUSCLE RELAXANTS

Whatever muscle-relaxation effect results from the use of these medications is probably due largely to a central nervous system depressive effect or mild analgesic effect; it is probably not worth the cost and the risk of adverse reactions. However, some patients who have taken these drugs in the past will request them, and other patients will respond better to (or at least be more receptive to the prescription of) one of the combination drugs listed in the second part of Table 3-VI than to simple analgesics.

If minor tranquilizers such as one of the benzodiazepines are considered for use as "muscle relaxants," it should be recognized that they probably also exert much of their action on the central nervous system, and the decision as to

Table 3-VI   Examples of Muscle Relaxants[a]

Drug alone	
Drug	Usual adult dosage
FLEXERIL (cyclobenzaprine HCl, Merck, Sharp & Dohme) 10 mg	i TID (maximum 2–3 weeks)
NORFLEX (orphenadrine citrate, Riker) 100 mg	i BID
PARAFLEX (chlorzoxazone, McNeil) 250 mg	i–ii TID–QID
RELA (carisoprodol, Schering) 350 mg	i QID
ROBAXIN (methocarbamol, Robins) 500 mg, 750 mg	6 gm/day acutely, then 4 gm/day (always in divided doses)
SKELAXIN (metaxalone, Robins) 400 mg	ii TID–QID
SOMA (carisoprodol, Wallace) 350 mg	i QID

Combinations with analgesics		
Drug	Ingredients	Usual adult dosage
NORGESIC (Riker)	Orphenadrine citrate 25 mg, aspirin 385 mg, caffeine 30 mg	i–ii TID–QID
NORGESIC FORTE (Riker)	Exactly double norgesic	½–i TID–QID
PARAFON FORTE (McNeil)	Chlorzoxazone 250 mg, acetamenophen 300 mg	ii QID
ROBAXISAL (Robins)	Methocarbamol 400 mg, aspirin 325 mg	ii QID
SOMA COMPOUND (Wallace)	Carisoprodol 200 mg, phenacetin 160 mg, caffeine 32 mg	i–ii QID

[a] This chart is for the reader's convenience only, and the listing may not be complete. Neither recommendation of the specific products listed nor criticism of those not listed is implied. Always check complete product information before prescribing any drug. All of these drugs can cause drowsiness.

whether to prescribe them should be based on a judgment as to whether their central nervous system effects are desirable in the individual case.

PSYCHOACTIVE MEDICATIONS AND RELAXATION TRAINING

Minor tranquilizers (sometimes in combination products with analgesics) can be useful in cases where anxiety is felt to be a factor in the initiation or the persistence of the patient's symptoms, or where the patient's reaction to the disability or to the prescribed therapy results in tension that complicates the situation. Their routine use in every patient with back or neck pain is to be condemned, however, and when they are prescribed it must be with a full awareness of the potential for interaction with alcohol and other drugs, for abuse, and for addiction.

Antidepressants are often helpful in patients with chronic neck and back pain, in whom depression is so often a contributing factor. The diagnosis and treatment (by pharmacologic or other means) of anxiety and depression is of course a branch of medicine in itself and is beyond the scope of this book.

The teaching of *relaxation techniques* can be the most important single therapeutic measure in patients with chronic muscle-tension states; some are discussed in Chapter 4. *Biofeedback training* is one method of teaching relaxation that seems to work well in a large number of patients and is becoming increasingly popular.

Patient Handout 3-1

Using Heat and Cold

Heat and cold are often very helpful measures that you can use at home to cut down on pain and spasm. If you have an acute injury, you should apply *cold* to the area for the first 48 hours or so. Otherwise, use whichever one (heat or cold) gives you the most relief. Often using *contrast* treatments (that is, heat immediately followed by cold or vice versa) is more beneficial than either one alone.

HEAT

If you ache or are stiff *all over*, you will find that a nice warm bath for 20 minutes or so first thing in the morning and before bed in the evening helps loosen your muscles and joints and relieves some of the pain.

If you hurt in a *specific area*, try applying heat to that area for twenty minutes or so four times a day. Never allow it to get so hot that it is painful or that the skin gets red. A heating pad is all right, but most people feel that *wet* heat is better. You can obtain wet heat in several ways:

By draping a towel over the sore area and letting the hot shower hit it (especially useful if it's your neck that hurts).

By soaking towels in hot water and applying them to the painful area. (The problem is that they tend to cool off quickly.)

By getting a heating pad specially made for wet heat. (*Never* allow a regular heating pad to get wet.)

By wrapping a wet towel around a hot water bottle.

By using a *hydrocollator,* which is a silicate gel pack that can retain heat for a prolonged period after being immersed in hot water. They are available in drug stores in a variety of shapes designed for specific body areas.

COLD

Never apply ice directly to the skin. If you don't have an ice bag, you can put some ice cubes in a plastic bag and then wrap the whole thing in a towel.

Apply the ice pack to the painful area for no more than 15 minutes at a time. If the pain is a long-term problem, do this four times a day; if you have just *injured* an area, do it every hour or two for the first 48 hours after injury.

Never use heat *or* cold over areas that are numb or that have poor circulation, or over open wounds. If you are diabetic, be sure to get your doctor's approval before starting.

4

The Neck and Upper Back

Conditions including *neurologic symptoms* and/or *diffuse pain in the upper extremities,* even if originating in the neck, are covered in Chapter 5. *Wryneck in infants and children* is discussed in Chapter 19. Further details on therapeutic suggestions can be found in Chapter 3.

For the anatomy of the cervical spine, see Figure 4-1.

EVALUATION

History

Inquire about the *duration* and *location* of pain. Recent sudden onset of unilateral pain and spasm without any known trauma is called ACUTE TOR-TICOLLIS.

Ask if there has been any *acute trauma or strain.* An acute strain of one area is

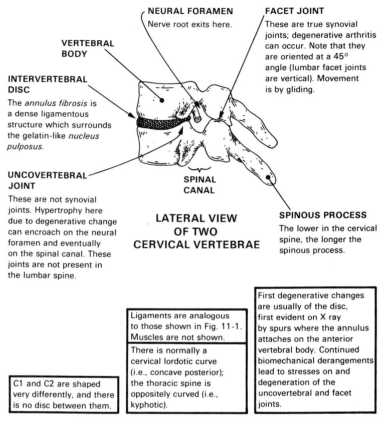

VERTEBRAL BODY

INTERVERTEBRAL DISC
The *annulus fibrosis* is a dense ligamentous structure which surrounds the gelatin-like *nucleus pulposus.*

NEURAL FORAMEN
Nerve root exits here.

FACET JOINT
These are true synovial joints; degenerative arthritis can occur. Note that they are oriented at a 45° angle (lumbar facet joints are vertical). Movement is by gliding.

UNCOVERTEBRAL JOINT
These are not synovial joints. Hypertrophy here due to degenerative change can encroach on the neural foramen and eventually on the spinal canal. These joints are not present in the lumbar spine.

SPINAL CANAL

LATERAL VIEW OF TWO CERVICAL VERTEBRAE

SPINOUS PROCESS
The lower in the cervical spine, the longer the spinous process.

C1 and C2 are shaped very differently, and there is no disc between them.

Ligaments are analogous to those shown in Fig. 11-1. Muscles are not shown.

There is normally a cervical lordotic curve (i.e., concave posterior); the thoracic spine is oppositely curved (i.e., kyphotic).

First degenerative changes are usually of the disc, first evident on X ray by spurs where the annulus attaches on the anterior vertebral body. Continued biomechanical derangements lead to stresses on and degeneration of the uncovertebral and facet joints.

Fig. 4-1. Anatomy of the cervical spine.

discussed on p. 38, whereas a CERVICAL SPRAIN refers to severe generalized pain such as in a whiplash injury.

Is there any *radiation* of pain down the arms, or *weakness, numbness,* or *paresthesias* in the arms? If so, see CERVICAL RADICULITIS in the next chapter.

If none of the above apply, see the general discussion of neck and upper back pain under the heading NECK PAIN SYNDROME.

Inquire about *precipitating factors*. Especially ask about *tension, stress,* and *postural factors* such as close work with the neck bent over, sleeping on too many pillows, or lying with the neck flexed while reading or watching TV.

Also ask about *relieving factors.*

Find out about *previous neck problems,* diagnoses, X rays, and treatments.

Examination

Look for *sites of tenderness*, especially over the spine; and also specifically look for *trigger points*, which are small spots of muscle spasm and are extremely tender.

Observe which *directions of motion* are restricted and/or painful: flexion (i.e., forward; normal 45°); extension (normal 55°); left and right rotation (normal 70° each way); and left and right lean (normal 40° each way).

Check *sensation, strength, reflexes,* and *pulses* in the arms.

NECK AND UPPER BACK PAIN

Acute Torticollis (Wryneck) (*Very Common*)

Diagnosis

There is a history of sudden onset (usually on awakening) of pain on one side of the neck with deviation of the neck to that side.

Often some premonitory twinges were noted before the acute event.

Frequently, localized exposure to cold (e.g., sleeping in a draft or sitting by an open car window) or prolonged positioning of the neck in an unusual position occurred just prior to the onset.

On examination, tenderness is usually diffuse on the involved side with palpable spasm, and motion away from the involved side is limited.

With repeated episodes, an X ray should be done to rule out underlying pathology.

The condition usually resolves in a few days.

Pathophysiology

Localized muscle spasm occurs due to muscle fatigue, environmental factors, or nerve irritation.

Treatment

A *soft cervical collar* for a few days helps to "unload" the tight muscle and support the neck in a neutral position.

Intermittent *heat* to the involved area is helpful in reducing pain and spasm.

Analgesics should be prescribed not only for temporary relief, but to help break the pain/spasm cycle.

Neck Pain Syndrome (*Extremely Common*)

Diagnosis

The pain may have been present for days to years, with different factors likely to be contributory depending on duration and other factors.
Often there is limited range of motion and crackling.
Especially look for trigger points on examination.

Pathophysiology

Various factors may be causing pain and stiffness alone or in combination:
muscle tension due to stress, anxiety, or fatigue
incorrect posture causing strain on soft tissues and muscle spasm
nerve root irritation from postural factors, degenerative changes, or muscle spasm; this is especially likely to be an important factor when pain is unilateral and fairly acute
degenerative disc and joint changes lead to pain through various mechanisms: stresses on pain-sensitive facet joints, on pain-sensitive soft-tissue structures, and by nerve root irritation. This is likely to be a significant factor in older patients and in those with chronic pain and stiffness. An X ray will show the extent of degenerative change
No matter what the inciting cause or causes (one or more of the above and/or an *acute strain*), all of these various factors can then interact and form vicious cycles:
pain → psychological tension → muscle tightness.
pain → muscle tightness directly.
pain → splinting → incorrect posture.
muscle tightness → incorrect posture → nerve root irritation.
degenerative changes → nerve root irritation.
nerve root irritation → muscle spasm and so on.
Prolonged limitation of motion leads to *loss of flexibility,* and therefore tissues which previously could tolerate certain positions and movements can no longer do so without pain.

Treatment

General Measures
Mild *analgesics* are helpful to break the pain cycle as well as for short-term relief.
A *soft cervical collar* is useful in an acute episode or on an intermittent basis,

but remember that prolonged use leads to loss of flexibility and muscle strength and may compound the problem.

Massage with *cold* or the application of *heat* can give some relief.

Specific Measures to be applied depending on history and findings (*any combination* may apply):

If trigger points are present, the most effective treatment is *injection* of those points. See Figure 4-2.

If *degenerative arthritis* is a factor, antiinflammatories should be tried.

If *postural problems* seem to be involved, the patient should be taught to avoid them. Education about neck care is not a bad idea in any case. See Patient Handout 4-2.

If *nerve root irritation* is a significant factor (especially likely when the problem is unilateral and fairly acute), antiinflammatory medication should

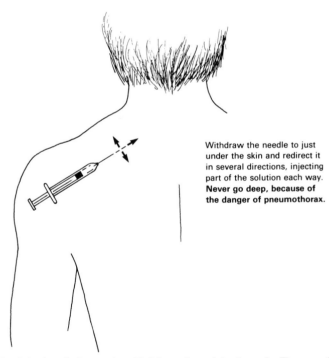

Withdraw the needle to just under the skin and redirect it in several directions, injecting part of the solution each way. **Never go deep, because of the danger of pneumothorax.**

Fig. 4-2. Injection of trigger points. Find the tender nodule of muscle. Then prep the area and spray with ethyl chloride. Using a *short 25-gauge needle,* inject into the muscle at the point of maximum tenderness at an oblique angle and not too deep. About one cc of 1% xylocaine without epinephrine is about right; some advocate mixing in some steroid, while others feel that even the xylocaine can be dispensed with, and that saline does just as well by physically breaking up the tight knot of muscle. See pp. 23-27 for further precautions and details before injecting.

Table 4-I   Specific Measures in Neck Pain Syndrome[a]

Contributing factor	Therapeutic measure
Trigger point	Injection (Fig. 4-2)
Degenerative arthritis	Antiinflammatory medications
Postural problems	Education (Patient Handout 4-2)
Nerve root irritation	Collar, antiinflammatory medications, traction prn (Patient Handout 5-1)
Psychogenic muscle tension	Various techniques (Patient Handout 4-3)
Loss of flexibility	Mobilization exercises (Patient Handout 4-4)
Loss of muscle tone	Strengthening exercises (Patient Handout 4-5)
Depression	As appropriate

[a] These are in addition to the general measures as described on pp. 35-36.

be used in addition to intermittent *heat* and a *soft collar*. If resolution does not occur, a trial of home *cervical traction* may be worthwhile. Technique for the latter is shown in Patient Handout 5-1. Traction is of course contraindicated in the presence of fracture, dislocation, rheumatoid arthritis or decreased bony strength (as from osteoporosis or metastatic disease).

If *psychogenic muscle tension* contributes and the problem is acute, a short course of an antianxiety agent may help. In chronic cases these are contraindicated because of the potential for addiction, and the patient should be encouraged to take up an appropriate aerobic exercise program, to get help with stress reduction and/or use a *muscle relaxation technique* (meditation, self-hypnosis, or the method discussed in Patient Handout 4-3).

In prolonged cases in which some *flexibility* has been *lost,* recommend the *mobilization exercises* shown in Patient Handout 4-4. These are contraindicated in patients with rheumatoid arthritis, acute injury, or decreased strength of bone.

If there is *loss of muscle tone, strengthening exercises* should be used (Patient Handout 4-5). These are contraindicated in patients with intracranial vascular disease or significant uncontrolled hypertension.

Because of the reinforcement and close follow-up that patients with chronic neck pain often need, it may sometimes be advantageous to refer them to a physical therapist or physiatrist.

A trial of *home cervical traction* may be considered in cases in which degenerative disc or joint changes are thought to play a role (see Patient Handout 5-1). Traction is of course contraindicated in the presence of fracture, dislocation, rheumatoid arthritis, or decreased bony strength (as from osteoporosis or metastatic disease).

Consider the possibility of a masked *depression* as the etiology of symptoms that are prolonged and not easily explained by the other factors discussed.

NECK AND UPPER BACK STRAIN

Acute Strain of Specific Neck or Upper Back Areas (*Very Common*)

Diagnosis

There is a history of acute strain, either with a sudden twist or prolonged abnormal posture.
Examination reveals tenderness localized in one area.

Treatment

Advise rest of the involved area (temporary use of a soft collar may be useful when neck muscles are involved).
Heat and *analgesics* should be prescribed.

Cervical Sprain (Whiplash) (*Common*)

Note: If the patient presents acutely, the neck must be immobilized in neutral position (with sandbags if necessary) while adequate X rays and neurologic examination are performed to rule out a fracture, dislocation, or spinal cord injury (any of which require continued immobilization and immediate neurosurgical consultation).

Diagnosis

Usually there is a history of forced flexion/extension of the neck.
The patient will commonly report that the onset of pain was not immediate after injury; often it begins the next day.
Diffuse tenderness is found.
The X ray is normal.
If there are any neurologic findings, the patient may have a ruptured cervical disc or a spinal cord injury, and should be referred to an orthopedist or neurosurgeon.

Treatment

Use a *soft collar* acutely (with progressively less use as time goes on to prevent stiffness).
Even *bedrest* may be necessary for a short time in severe cases.

Ice application is useful in the first 48 hours; after that *wet heat* may be more beneficial.

Sufficient *analgesics* should be prescribed; *antiinflammatory drugs* may also be helpful.

As time goes on the various *other factors* discussed in the section on NECK PAIN SYNDROME may become operative and should be addressed as discussed there (for example: psychogenic muscle tension, loss of flexibility, loss of muscle strength).

Referral to a physical therapist or physiatrist for follow-up should be considered in cases that are at all severe or prolonged.

Be aware that *secondary gain* issues (financial and otherwise) are often involved in these cases.

Patient Handout 4-1

Using a Cervical Collar

A collar can be extremely helpful in the treatment of various neck problems, but only if it is worn properly. Be sure to follow these guidelines:

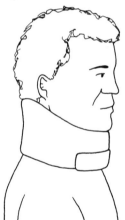

1. Make certain that it is the proper size to hold your head straight as in the picture, not tipped back or forward.

2. If you have been given the collar because of a recent strain or wryneck, leave it on for most or all of the day at first; after a couple of days begin wearing it less and less each day.

3. If your neck problem is long term, be sure to follow your doctor's advice as to how often and for how long to wear your collar.

Patient Handout 4-2

Taking Care of your Neck

Here are some tips on preventing strain to your neck. In addition to these measures, you should also do whatever else your doctor has recommended for your specific neck problem.

1. SOME GENERAL RULES

WRONG   WRONG

Avoid tilting your head backward or forward for a prolonged period.
Don't turn or lean your head far to one side.
Never hold your neck in one position for a long time. Move it around every so
often.

WRONG

Don't sleep on your stomach, as your neck will be twisted to one side.

RIGHT

If you sleep on your side, be sure your pillow is just thick enough to hold your head neutral, tilted neither up nor down.

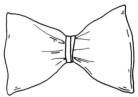

"BUTTERFLY" PILLOW

If you sleep on your back, use a "butterfly" pillow under your neck. (You can make one by tying a ribbon around the center of a soft pillow.)

3. SITTING

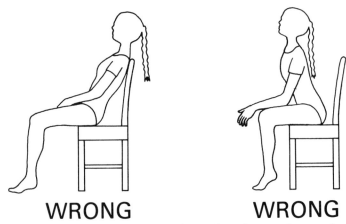

WRONG   WRONG

Don't slouch your shoulders or lean forward so that you have to look up to compensate.

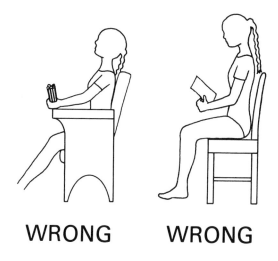

WRONG   WRONG

Don't have your work or your reading material so low (in your lap, for example) or so high that you have to sit with your head bent over or tilted up.

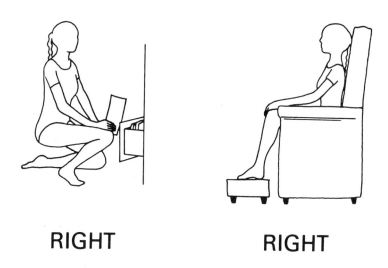

RIGHT   RIGHT

Support your head with a high-backed chair while watching TV.

4. STANDING

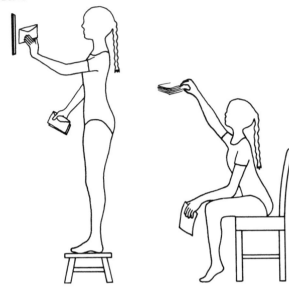

RIGHT   WRONG

Use a stepladder or stoop to keep your work at eye level, so that you don't have to look continually up or down.

5. DRIVING

Don't drive with your chin up. Make sure your head restraint is adjusted to meet the back of your head, not your neck. Always wear your seat and shoulder belts.

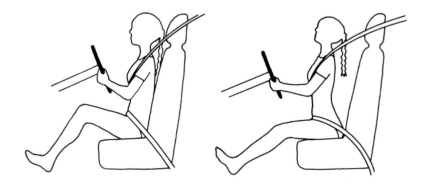

RIGHT   WRONG

6. SHAVING OR PUTTING ON MAKEUP

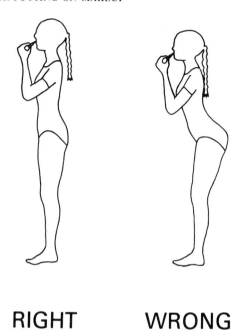

RIGHT   WRONG

Don't lean forward over the sink, forcing yourself to raise your chin.

7. DRINKING

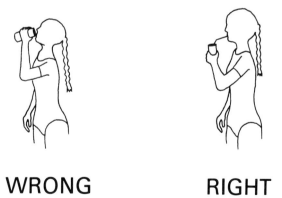

WRONG   RIGHT

Don't drink out of a bottle or can, and don't empty a glass to the very last drop. Using a straw is helpful.

8. USING THE PHONE

WRONG   RIGHT

Don't cradle the phone between your head and shoulder.

9. WEARING BIFOCALS

WRONG   RIGHT

If you constantly have to hold your chin up to read or do close work, get a separate pair of reading glasses instead of using bifocals.

10. AT THE MOVIES

Don't sit too close to the front where you will have to be looking up.

Patient Handout 4-3

Relaxation

When you are tense, your muscles become tense, especially those in your neck, head, and back. After muscles have been tight for a while, they begin to hurt. If muscles are already sore from bad posture, a pinched nerve or whatever, those muscles will be the first to tighten up when you get tense. So one of the most important things you can do to help relieve pain coming from muscles in your neck, back or head is to learn to relax.

There are a variety of techniques to do this. Some are called "mediation," some are called "self-hypnosis" and others have no specific name. You may wish to get a good book on one of these topics, such as *The Relaxation Response* by Herbert Benson, M.D. Taking up regular aerobic exercise such as walking, running, cycling, or swimming can also be extremely helpful. (Get your doctor's approval before starting.)

One relaxation technique is printed below. You should go through it at least twice a day to start; hopefully you will eventually incorporate relaxation into your approach toward life, or at least be able to grab hold of yourself and relax when stresses begin to pile up.

1. Lie down on your back, or sit in a comfortable chair. Close your eyes.

2. Breathe in slowly for a count of two and then out for a count of two. Keep breathing in this slow rhythm.

3. Make a fist in both hands and then relax them. Feel the relaxation spread up your arms and into your neck, and then down to your back and legs. Keep up the slow breathing.

4. Make a conscious effort to exclude from your mind anything but an awareness of your slow, rhythmic breathing, and thinking how very, very relaxed you are. Think about each part of your body in turn, and how relaxed it is; then think about your breathing; then think about each part of your body and how relaxed it is again.

5. Keep this up for at least 10 minutes.

Patient Handout 4-4

Neck Mobility Exercises

If your neck has lost flexibility, that is, become stiff, these exercises can help restore a normal range of motion. Start all exercises with your neck in a neutral position. If there is pain, stop doing the exercises and let your doctor know.

1. FLEXION

Lean your head forward as far as it will go *without forcing* it, count to five, and return to straight up. Don't tilt it backwards. Do five repetitions, two or three times a day.

2. SIDE LEANING

Tilt your head to the left without twisting it, as far as it will go *without forcing* it, and count to five. Now straighten it and do the same thing the other way. Do five repetitions each way, two or three times a day.

3. ROTATION

Twist your head to the left as far as it will go *without forcing* it, and count to five. Now turn back forward and do the same thing to the right. Do five repetitions each way, two or three times a day.

Patient Handout 4-5

Neck Strengthening Exercises

Restoring strength to the muscles of your neck can help support the other tissues so that they hurt less. Do all exercises with your head in a neutral position; try to relax completely between repetitions and to breathe normally.

1. FLEXION

Put your hands on your forehead and push your head hard into your hands; your hands push back and prevent any movement. Hold for a count of five and relax. Start with three repetitions twice a day and slowly increase.

2. EXTENSION

Put your hands behind your head and push your head backwards; your hands push back and prevent any movement. Hold for a count of five and relax. Start with five repetitions twice a day and slowly increase.

3. LEANING

Put your hand on the side of your head and push your head into it; your hand pushes back and prevents any movement. Hold for a count of five and relax. Start with three repetitions each way, twice a day, and slowly increase.

4. ROTATION

Put your right hand on the right side of your forehead and your left hand on the left side of the back of your head. Now try to twist your head to the right while your hands prevent any movement. Hold for a count of five and relax. Do the same thing the other way, with hands reversed. Start with three repetitions each way twice a day, and slowly increase.

5

Neurologic Symptoms and Diffuse Pain of the Upper Extremity

Further details on some therapeutic suggestions can be found in Chapter 3.

EVALUATION

Pain and neurologic symptoms of the upper extremity can be due to *vascular compromise* by emboli, arteriosclerosis, Raynaud's phenomenon, etc. The discussion of these conditions is beyond the scope of this book, but their possibility must not be forgotten; appropriate history must be taken and necessary examination and tests performed in order to rule them out.

Also, weakness, numbness, or paresthesias obviously are often due to lesions of the *brain* or *spinal cord*. A *polyneuropathy* will usually manifest itself as sensory loss in a "glove" distribution without associated pain, but metabolic disease (especially *diabetes*) can affect single large nerves and lead to confusion with the compression neuropathies discussed herein. And diseases of *muscle* or the *neuromuscular junction* can underlie motor weakness.

Again, discussion of these diseases is beyond the scope of this volume; but it is imperative that further history, examination, and testing—often well beyond that which is discussed below—be undertaken to rule out such an etiology whenever the original history and/or examination is at all suggestive. These diseases usually will not have significant pain associated with the neurologic symptoms or findings, but this is not a completely reliable clue. And do not forget that an arm ache can be a manifestation (sometimes the only one) of serious *intrathoracic disease* such as angina pectoris, an acute myocardial infarction or an apical pulmonary tumor. Index of suspicion must remain high, and threshhold for doing ECG, chest X ray etc. kept low.

History

Ask about the *duration* of symptoms and the *location* of *pain* and *paresthesias*. Often the answer to the latter question is too vague to be useful. See Figure 5-1. A history of nonspecific aching in both shoulders and upper arms in an older patient without other specific history or findings is suggestive of POLYMYALGIA RHEUMATICA, discussed in Chapter 21.

Has the patient noticed any *weakness* or *incoordination*?

Inquire about precipitating factors:

Pain mainly at night, especially after much wrist usage during the day, is suggestive of the CARPAL TUNNEL SYNDROME.

Paresthesias and pain which come on with the arm in certain positions (especially while sleeping or with the arm leaning on a hard object) are suggestive of a THORACIC OUTLET SYNDROME or peripheral nerve compression.

Pain brought on by general arm usage and relieved with rest should bring to mind the possibility of vascular disease.

Continuous symptoms, especially if pain is not a significant factor, must lead to consideration of a CNS lesion.

Pain brought on by exposure to cold, especially if associated with color changes, raises the possibility of *Raynaud's Phenomenon*.

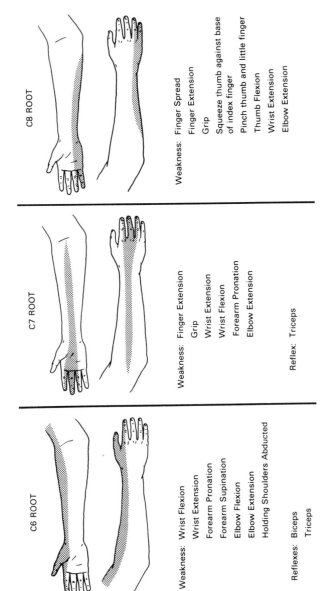

C6 ROOT

Weakness: Wrist Flexion
Wrist Extension
Forearm Pronation
Forearm Supination
Elbow Flexion
Elbow Extension
Holding Shoulders Abducted

Reflexes: Biceps
Triceps

C7 ROOT

Weakness: Finger Extension
Grip
Wrist Extension
Wrist Flexion
Forearm Pronation
Elbow Extension

Reflex: Triceps

C8 ROOT

Weakness: Finger Spread
Finger Extension
Grip
Squeeze thumb against base
of index finger
Pinch thumb and little finger
Thumb Flexion
Wrist Extension
Elbow Extension

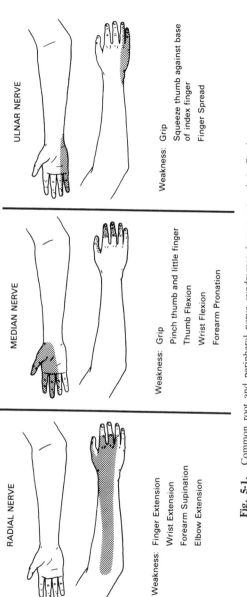

RADIAL NERVE

Weakness: Finger Extension
Wrist Extension
Forearm Supination
Elbow Extension

MEDIAN NERVE

Weakness: Grip
Pinch thumb and little finger
Thumb Flexion
Wrist Flexion
Forearm Pronation

ULNAR NERVE

Weakness: Grip
Squeeze thumb against base
of index finger
Finger Spread

Fig. 5-1. Common root and peripheral nerve syndromes (upper extremity). Syndromes are usually incomplete. Dermatomes and innervations shown can vary in some patients.

Intermittent numbness and tingling in both arms, especially if brought on by stress or associated with perioral numbness, is suggestive of *hyperventilation*. See if the symptoms can be reproduced with rapid deep breathing in the office.

Has there been any *neck pain* or recent *neck trauma*? These would suggest CERVICAL RADICULITIS.

Question the patient about any *weakness, numbness, paresthesias* or *incoordination* elsewhere in the body, or any headaches, visual or auditory symptoms, vertigo, imbalance, speech difficulty, episodes of loss of consciousness, or recent head trauma, any of which may suggest a CNS lesion.

Examination

Examine the *shoulder, elbow,* and *wrist* for *range of motion, swelling,* and *tenderness,* since often a specific articular or periarticular syndrome (e.g., shoulder bursitis, epicondylitis) will be the cause of the patient's complaints even though he or she is unable to localize the symptoms.

Examine the *neck* and paracervical area for tenderness and check range of motion of the cervical spine.

Check distal *pulses* and *capillary filling.*

Elicit the biceps or brachioradialis and the triceps *reflexes* (Figure 5-2).

Test muscle *strength.* See Table 5-1 for the various motions to be tested and their root and peripheral innervations. Significant weakness is always an indication for referral for more precise diagnosis (e.g., EMG) and more aggressive management.

Look for muscle *atrophy* (especially that seen in the thenar eminence with severe or prolonged CARPAL TUNNEL SYNDROME).

Test for pin *sensation.* When a more sensitive examination needs to be done, check two-point discrimination. See Figure 5-1 for the areas classically involved with the various cervical roots and peripheral nerves; but note that areas are often

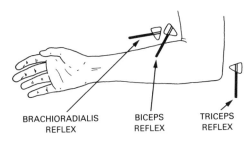

BRACHIORADIALIS   BICEPS   TRICEPS
REFLEX   REFLEX   REFLEX

Fig. 5-2.   Reflexes: upper extremity.

Table 5-I   Outline for Examination of Patients with Diffuse Pain and/or Neurologic Symptoms
in the Upper Extremity[a]

1. Check for tenderness around, and range of motion of, the wrist, elbow and shoulder
2. Check for tenderness and range of motion of the neck
3. Check vascular status: radial and ulnar pulses, distal skin color and temperature, and capillary
   filling (press the nail and see how long it takes the color to return to the nailbed)
4. Check the biceps or brachioradialis (C5–6) and the triceps (C6–7) reflexes
5. Test muscle strength by comparing with the uninvolved side (all maneuvers to be done against
   your resistance; root and peripheral innervations are as noted below)
   FINGER SPREAD (C8,T1; ulnar)
   HOLDING FINGERS EXTENDED AT MCP AND PIP JOINTS (C7,8; radial)
   GRIP (RADIAL AND ULNAR SIDES) (C7–T1; median and ulnar)
   SQUEEZE THUMB AGAINST BASE OF INDEX FINGER (C8,T1; ulnar)
   PINCH BETWEEN THUMB AND LITTLE FINGER (C8,T1; median)
   FLEX THUMB (C8; median)
   EXTEND WRIST (C6–8; radial)
   FLEX WRIST (C6–7; median)
   SUPINATE FOREARM WITH ELBOW STRAIGHT (C5–6; radial)
   PRONATE FOREARM WITH ELBOW STRAIGHT (C6–7; median)
   FLEX ELBOW WITH FOREARM SUPINATED (C5–6; musculocutaneous)
   STRAIGHTEN ELBOW (C6–8; radial)
   HOLD SHOULDERS ABDUCTED (C5–6; axillary)
6. Screen pin sensation in the following areas (innervations shown):
   WEB SPACE BETWEEN THUMB AND INDEX FINGER ON THE BACK OF THE HAND
   (C6; radial)
   PALMAR SIDE OF THE DIP JOINT OF THE MIDDLE FINGER (C7; median)
   SIDE OF THE HAND PROXIMAL TO THE FIFTH DIGIT (C8; ulnar)
   SIDE OF THE MID-FOREARM PROXIMAL TO THE FIFTH DIGIT (T1)
7. Perform the specific maneuvers shown in Figure 5-3 in sequence

[a] With some practice the entire examination will take no more than a few minutes. Of course,
additional examination should be done as indicated.

incompletely involved, that sensation testing is subjective and thus often quite
inaccurate unless the deficit is almost complete, and that there are individual
variations from the dermatomes shown.

Do the following maneuvers (Figure 5-3); reproduction or intensification of
symptoms helps to define the problem as listed:

Full flexion of the neck: CERVICAL RADICULITIS on a discogenic basis
The foramenal compression test (Spurling's maneuver): CERVICAL RADI-
CULITIS from nerve root impingement in the neural foramen
The thoracic outlet maneuvers:
Adson's maneuver: the ANTERIOR SCALENE SYNDROME
Costoclavicular maneuver: the COSTOCLAVICULAR SYNDROME

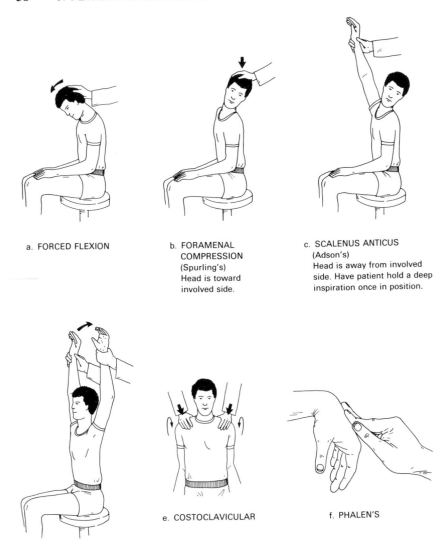

a. FORCED FLEXION

b. FORAMENAL
COMPRESSION
(Spurling's)
Head is toward
involved side.

c. SCALENUS ANTICUS
(Adson's)
Head is away from involved
side. Have patient hold a deep
inspiration once in position.

e. COSTOCLAVICULAR

f. PHALEN'S

d. HYPERABDUCTION
(Wright's)

Fig. 5-3. Maneuvers. Each should be maintained for at least 10 seconds, except Phalen's maneuver, which should be held for 30 seconds. Reproduction of distal symptoms is a positive result.

Hyperabduction maneuver: the PECTORALIS MINOR SYNDROME
Phalen's maneuver: the CARPAL TUNNEL SYNDROME

Electrodiagnosis

When the diagnosis cannot be pinpointed by the history and examination described above, referral should be made for an EMG/NCV (described in Chapter 2).

X Ray

A cervical spine X ray and a chest X ray should be obtained in every patient with objective neurologic findings not clearly due to a peripheral problem such as CARPAL TUNNEL SYNDROME. Referral for myelography is warranted in cases with severe or progressive deficit or if there is suspicion of a spinal cord lesion.

SYNDROMES

Cervical Radiculitis (''Pinched Nerve'') (*Very Common*)

Diagnosis

Almost invariably there is pain and tenderness in the neck as well as distal symptoms.

If pain, paresthesias, and numbness can be localized, then the pattern is consistent with one or more cervical dermatomes.

Any muscle weakness and hyporeflexia is consistent with the involved roots. (Remember that significant or progressive weakness is an indication for referral for further investigation to rule out tumor, etc.) See Figure 5-1.

Neck flexion (Figure 5-3a) reproduces or aggravates the distal symptoms when root irritation is by a disc.

The foramenal compression test (Figure 5-3b) causes distal symptoms if degenerative arthritis with foramenal encroachment is responsible.

An X ray should be done, and usually will show evidence of degenerative arthritis, foramenal encroachment, and/or disc disease at the involved levels.

Pathophysiology

Impingement on one or more nerve roots is by a bulging or herniated disc, or by narrowing of the exit foramena, or from degenerative hypertrophy of the posterior joints.

Inflammation and swelling around the root may be an initiating or a contributing factor as the root travels in a previously narrowed but not yet impinging pathway.

Treatment

Use an *antiinflammatory medication.*

A *soft cervical collar* should be worn, with its use tapered as the patient responds (Patient Handout 4-1).

Wet *heat* to the neck may be helpful.

If these measures do not bring improvement, a trial of home *cervical traction* should be considered (Patient Handout 5-1.) Traction is contraindicated in rheumatoid arthritis, acute trauma, instability, and conditions causing decreased strength of bone.

Cases with significant or progressive motor weakness, or which are not responding to these measures, should be referred to the specialist.

Thoracic Outlet Syndromes (*Uncommon*)

Diagnosis

Diffuse pain and paresthesias are usually positional.

Symptoms are reproduced by Adson's maneuver (SCALENUS ANTICUS SYNDROME), the costoclavicular maneuver (the COSTOCLAVICULAR SYNDROME), or by the hyperabduction maneuver (the PECTORALIS MINOR SYNDROME) (Figure 5-3).

X ray may show a cervical rib, but this is probably not often the cause of the patient's symptoms. Chest X ray should be done to rule out an apical chest lesion.

Pathophysiology

Compression of the neurovascular bundle occurs as it passes through and around various structures: between the scalenus anticus muscle and the first rib, between the first rib and the clavicle, or between the pectoralis minor muscle and the first rib.

Stress, poor posture, and muscular fatigue are usually what allow narrowing at these various locations.

Treatment

In patients with intermittent symptoms, *avoidance* of the responsible position is all that is necessary.

Patients with more persistent symptoms should be advised to practice consciously not allowing the shoulders to droop and to follow the precautions itemized in Patient Handout 4-2.

Surgery to remove a cervical rib is rarely necessary.

Brachial Plexitis (*Rare*)

Diagnosis

The patient presents with fairly acute, diffuse, and usually severe axillary, arm and hand pain, paresthesias and often weakness, without known trauma.

Especially look for proximal muscle atrophy and scapular winging.

There is usually axillary tenderness on examination.

Cervical spine and chest X rays must be done to rule out lesions there.

Blood sugars should be done to rule out a diabetic neuropathy.

Pathophysiology

Idiopathic inflammation of the brachial plexus is responsible, possibly on a postviral autoimmune basis.

Treatment

Place the arm in a *sling*.

Potent *antiinflammatory medication* should be used.

The problem sometimes resolves in a week or two. Referral to a neurologist is wise if the patient does not improve fairly quickly.

Carpal Tunnel Syndrome (Median Nerve Compression in the Wrist) (*Common*)

Diagnosis

The patient complains of aching pain in the first three digits and radial side of the hand, often with paresthesias and numbness. Note that the pain is often more diffuse and can extend well up the arm.

Symptoms are usually predominant at night and are more pronounced following days in which the wrist has been overused (common in carpenters and supermarket checkers, for example.)

On examination there may be decreased sensation in the median nerve distribution (Figure 5-1).

Phalen's maneuver should produce aching pain and paresthesias (Figure 5-3f).

Tapping the middle of the volar wrist may more easily produce distal paresthesia than in the uninvolved extremity (Tinel's sign).

Check for atrophy of the thenar eminence or objective weakness (best evaluated by testing the strength of opposition of the tip of the thumb to the tip of the little finger, and of extension of the second and third digits at the PIP joint with the MCP joint hyperextended).

The condition is common in pregnancy, and also found in diabetes mellitus, acromegaly, rheumatoid arthritis, and myxedema. The latter four conditions should be tested for in severe, prolonged or otherwise unexplained cases, and X rays to rule out impinging bony lesions should be ordered as well.

Pathophysiology

Swelling in the carpal tunnel of the wrist causes compression of the median nerve. Rarely, a bony lesion or soft-tissue mass is responsible for the compression.

Treatment

A cock-up *wrist splint* (Figure 8-5) should be used during the day and especially during offending activity, even if most of the symptomatology is nocturnal.

Oral *antiinflammatory medications* should be prescribed.

If these measures are unsuccessful, *injection* of steroid should be offered. Success rate is good. See Figure 5-4.

If that is unsuccessful, or if at any time there is found thenar atrophy or motor weakness on examination, the patient should be referred for confirmatory NCV testing and then *surgery* (in which the transverse carpal ligament is incised).

Pronator Teres Syndrome (Median Nerve Compression in the Forearm) (*Rare*)

Diagnosis

Usually a history is given of direct trauma or compression of the proximal forearm or excessive pronation/supination movement.

Symptoms and neurologic exam are analogous to the CARPAL TUNNEL SYNDROME. (Differences are usually too subtle to detect.)

Phalen's test is negative.

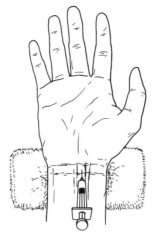

Fig. 5-4. Carpal Tunnel Injection. Have the patient hold his wrist slightly dorsiflexed over the edge of the treatment table or over a rolled towel. Find the distal transverse crease and inect just ulnar to the prominent palmaris longus tendon, with the needle pointing at a 45° angle toward the hand. Use a 25-gauge needle and inject about a ½ cc. of steroid and ½ cc. of xylocaine without epinephrine. See p. 23 for further details and precautions before injecting.

Full pronation of the forearm intensifies the symptoms.
There is often tenderness to deep palpation of the proximal volar forearm.

Pathophysiology

The median nerve is compressed as it passes between the two heads of the pronator teres muscle.

Treatment

Have the patient *avoid* the precipitating compression or activity.
If weakness or atrophy develops, referral should be made for electrodiagnostic confirmation and *surgery*.

Tardy Ulnar Palsy (Cubital Tunnel Syndrome: Ulnar Nerve Compression at the Elbow) (*Uncommon*)

Diagnosis

Found in patients who lean on their elbow or who have traumatized their elbow.
Paresthesias and possibly decreased sensation occur in the fourth and fifth digits.

Pressure on the cubital tunnel (between the olecranon and the medial epicondyle) reproduces or intensifies the symptoms.

Froment's sign is classic. The harder the patient tries to pinch between thumb and index finger, the more the distal phalanx of the thumb flexes.

When severe, weakness of finger spread and interosseus atrophy can be found.

Pathophysiology

The ulnar nerve suffers acute or repetitive injury in the cubital tunnel.

Treatment

Use *padding* and/or *avoidance* of the responsible position.

Refer for electrodiagnostic confirmation and *surgery* when symptoms are prolonged or severe, or if there is motor weakness or atrophy.

Ulnar Nerve Compression at the Wrist (*Uncommon*)

Diagnosis

Symptoms and neurologic findings are as in a TARDY ULNAR PALSY.

Palpation of the pisiform-hamate tunnel may cause symptoms.

Pathophysiology

Compression of the ulnar nerve can occur where it passes through the "canal of Guyon" between the pisiform and hamate bones or can be secondary to trauma or to bony or soft tissue anomolies.

Treatment

If there are significant sensory symptoms or any motor findings, the patient should be referred for electrodiagnostic confirmation and *surgical release*.

Radial Nerve Palsy (*Rare*)

Diagnosis

An acute radial nerve palsy will be obvious (wrist drop, impaired sensation) and follows prolonged compression, usually during unconsciousness.

More chronic symptoms can be produced by compression of the nerve as it passes in the area around the lateral epicondyle.

Treatment

Referral for management by the specialist is appropriate.

Patient Handout 5-1

Home Neck Traction

This is a technique that can often be very helpful in relieving symptoms from a pinched nerve in the neck.

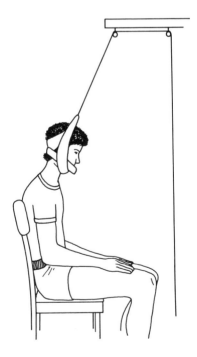

1. Follow any specific instructions that your doctor gives you, even if they are different from what is written here.
2. Sit facing the door where you have installed the pulleys. The head harness should fit around the base of your skull and under the point of your chin. Sit far enough away so that it pulls at about a fifteen degree angle.
3. Fill the bag with 5 lb of water, and slowly let it off your lap so that there isn't a sudden jerk.
4. Always have someone within yelling distance.
5. Continue the traction for 20 minutes or so; do it twice a day.
6. If the traction produces pain or other symptoms, stop doing it until you have discussed it with your doctor.

6

The Shoulder

Pain in the *neck muscles* is discussed in Chapter 4. Further details on therapeutic suggestions made in this chapter can be found in Chapter 3.

For the anatomy of the shoulder, see Figure 6-1.

SHOULDER PAIN AND LIMITATION OF MOTION

Note: Always consider the possibility of shoulder pain being *referred* from *intrathoracic disease* or diaphragmatic irritation (ischemic heart disease, pulmonary tumors, and gall bladder disease, for example). While the differential diagnosis and management of such conditions is obviously beyond the scope of this book, they must be kept in mind and appropriate history, examination, and tests done to rule them out, especially if the initial history and/or examination is not completely consistent with a musculoskeletal origin of the patient's pain.

History

Ask about the *duration* and *location* of the pain. (If the patient is actually referring to the posterior musculature rather than the shoulder area itself, refer to Chapter 4 on neck pain.)

What are *precipitating* and *relieving* factors?

Has there been any specific *strain, overuse,* or *trauma*?

Inquire about any *neck pain* or *trauma, radiation* of pain into the arms or neurologic symptoms in the arms. (Again, if any are present see Chapter 4 on neck pain or the discussion of radiated arm pain in Chapter 5.)

Ask about *previous shoulder problems,* diagnoses, and treatments.

Are any *other joints* involved? Does the patient have a history of arthritis or gout?

Examination

Look for *tenderness* (see Figure 6-2).

Test *range of motion* (see Figure 6-3).

Limitation of all motions implies ADHESIVE CAPSULITIS or much more rarely DEGENERATIVE ARTHRITIS OF THE SHOULDER.

Limitation (due to pain) of elevation only, or elevation and one or two other motions, indicates ROTATOR CUFF TENDONITIS or SUBACRO-MIAL BURSITIS (but is also consistent with ACROMIOCLAVICULAR JOINT ARTHRITIS if tenderness was found there).

Loss of the ability to elevate the shoulder actively past a certain point, not secondary to pain, is indicative of a ROTATOR CUFF TEAR or muscular weakness.

Look for *swelling, redness,* or *warmth* which would imply an acute inflammatory process.

Check *distal strength, sensation,* and *pulses.*

Rotator Cuff Tendonitis and Subacromial Bursitis (*Very Common*)

Diagnosis

The symptoms may be very acute, very chronic or anywhere in between. (The so-called *hyperacute* shoulder is probably secondary to the release of calcium crystals into the bursa.)

A history of specific overuse (usually with the arm above neck level) is more

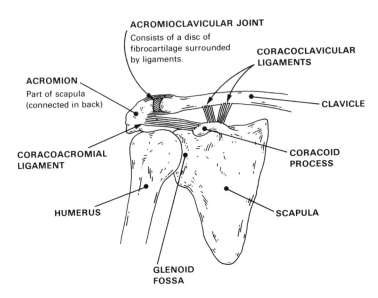

ACROMIOCLAVICULAR JOINT
Consists of a disc of fibrocartilage surrounded by ligaments.

CORACOCLAVICULAR LIGAMENTS

ACROMION
Part of scapula (connected in back)

CLAVICLE

CORACOACROMIAL LIGAMENT

CORACOID PROCESS

HUMERUS

SCAPULA

GLENOID FOSSA

ANTERIOR VIEW: BONES AND LIGAMENTS

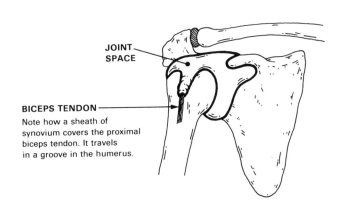

JOINT SPACE

BICEPS TENDON
Note how a sheath of synovium covers the proximal biceps tendon. It travels in a groove in the humerus.

ANTERIOR VIEW: JOINT SPACE

Fig. 6-1. Anatomy of the shoulder.

often elicited when the patient is younger and tendonitis is the predominant finding.

The pain may be referred distally to the area of the deltoid muscle insertion.

You may find subacromial tenderness (when tendonitis is predominant) or no localizable tenderness (sometimes when bursitis is predominant).

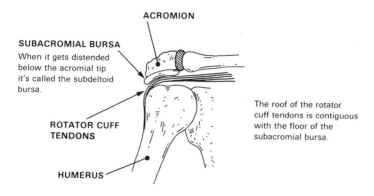

ACROMION

SUBACROMIAL BURSA
When it gets distended
below the acromial tip
it's called the subdeltoid
bursa.

The roof of the rotator
cuff tendons is contiguous
with the floor of the
subacromial bursa.

**ROTATOR CUFF
TENDONS**

HUMERUS

**ANTERIOR VIEW: CUFF TENDONS AND
SUBACROMIAL BURSA**

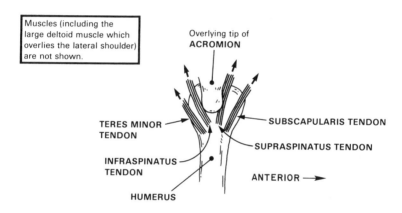

Muscles (including the
large deltoid muscle which
overlies the lateral shoulder)
are not shown.

Overlying tip of
ACROMION

**TERES MINOR
TENDON**

SUBSCAPULARIS TENDON

SUPRASPINATUS TENDON

**INFRASPINATUS
TENDON**

ANTERIOR ➤

HUMERUS

**LATERAL VIEW:
ROTATOR CUFF TENDONS**

Fig. 6-1. Continued

Elevation is painful past a certain point (anywhere from 60° to 105°).
Sometimes other motions (but usually not all) are painful as well.
The X ray may reveal calcium in the bursa or along the tendon in long-standing
cases.

Pathophysiology

The blood supply of the cuff tendons is quite tenuous in the area near the
acromial tip. The interaction between this propensity to ischemia and mechanical

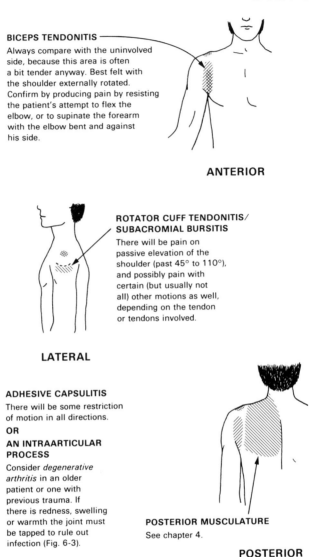

BICEPS TENDONITIS
Always compare with the uninvolved
side, because this area is often
a bit tender anyway. Best felt with
the shoulder externally rotated.
Confirm by producing pain by resisting
the patient's attempt to flex the
elbow, or to supinate the forearm
with the elbow bent and against
his side.

ANTERIOR

**ROTATOR CUFF TENDONITIS/
SUBACROMIAL BURSITIS**
There will be pain on
passive elevation of the
shoulder (past 45° to 110°),
and possibly pain with
certain (but usually not
all) other motions as well,
depending on the tendon
or tendons involved.

LATERAL

ADHESIVE CAPSULITIS
There will be some restriction
of motion in all directions.
OR
**AN INTRAARTICULAR
PROCESS**
Consider *degenerative
arthritis* in an older
patient or one with
previous trauma. If
there is redness, swelling
or warmth the joint must
be tapped to rule out
infection (Fig. 6-3).

POSTERIOR MUSCULATURE
See chapter 4.

POSTERIOR

AC JOINT ARTHRITIS
Extreme elevation will
be painful.

Fig. 6-2.   Sites of tenderness: the shoulder.

compression by the overlying acromion (as the shoulder is elevated) leads to
inflammation and degenerative changes.

Other motions can be involved; for example, swimmers very commonly de-

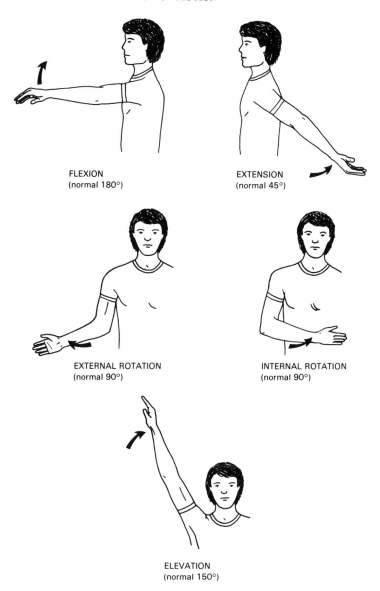

FLEXION
(normal 180°)

EXTENSION
(normal 45°)

EXTERNAL ROTATION
(normal 90°)

INTERNAL ROTATION
(normal 90°)

ELEVATION
(normal 150°)

Fig. 6-3. Shoulder range of motion.

velop the syndrome secondary to impingement by the coracoacromial ligament on the anterior rotator cuff.

The bursa is usually inflamed secondary to inflammation in the rotator cuff; the floor of the bursa is inseparable from the superior aspect of the rotator cuff.

Treatment

Advise *rest* and *avoidance* of the responsible activity.

However, in cases that drag on for weeks, especially in older patients, it is imperative that *active range of motion* be maintained (see Patient Handout 6-1 for a series of exercises to prevent the development of ADHESIVE CAPSULITIS; do not have the patient do the sideways wall climbing).

Ice is helpful in acute cases, *heat* in more chronic ones.

Oral *antiinflammatories* may be used.

Steroid injection has a very good success rate and is considered by many to be the treatment of choice, especially in the hyperacute situation. See Figure 6-4.

In long-term cases, lack of response may be due to the presence of a partial rotator cuff tear, either degenerative in nature or secondary to unrecognized trauma. Referral for arthrography will be necessary to rule this out.

In the absence of a tear, *surgery* should be considered a last resort, though sometimes removal of calcium deposits or acromionectomy does give good results.

Biceps Tendonitis (*Common*)

Diagnosis

Tenderness is found over the biceps tendon with the shoulder externally rotated. Always compare to the uninvolved side.

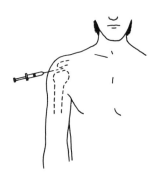

Fig. 6-4.   Injection for cuff tendonitis and subacromial bursitis. Prep the area and spray with ethyl chloride. Use a 23-gauge needle. If there is subacromial tenderness, inject about ½ cc of steroid and ½ cc of local anesthetic around (never into) the sore tendon. If no tenderness is found, omit this and proceed directly to the second step, which is the injection of a cc each of steroid and anesthetic subacromially as shown. The patient must be told to avoid ballistic motions (such as throwing a ball or swinging a racket) for 3 weeks or so. Injections may be repeated at weekly intervals if necessary up to a total of three in a course. See pp. 23–27 for further details and precautions.

The diagnosis is confirmed by reproducing pain there with resisted elbow flexion and/or resisted forearm supination with the elbow bent.

Pathophysiology

Inflammation and degenerative changes of the tendon occur due to overuse. Sometimes the tendon pops out of its groove recurrently (the most common cause of the "snapping" or "popping" shoulder).

Treatment

Use *rest, heat,* and *antiinflammatories.*
Injection may be offered. Success rate is good. See Figure 6-5.
Note that degenerative changes in the tendon can predispose to *rupture* with the characteristic appearance of a bulge in the distal upper arm. Repair is necessary only if the patient requires powerful supination of his forearm at work or play, or for cosmetic reasons. If it is to be done, however, it must be done soon after rupture.

Acromioclavicular Joint Arthritis (*Uncommon*)

Diagnosis

Tenderness is found at the acromioclavicular joint (see Figure 6-2). Pain occurs with high elevation of the shoulder.

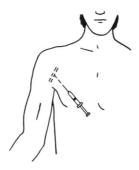

Fig. 6-5. Injection for bicipital tendonitis. Have the patient hold the arm externally rotated. Find the tender biceps tendon and after prepping the area and spraying with ethyl chloride, use a 23-gauge needle to inject a cc of steroid and a cc of local anesthetic around (never into) the sore tendon. The patient must be told to avoid ballistic motions for at least 3 or 4 weeks. See pp. 23–27 for further details and precautions.

Pathophysiology

Degenerative changes in the fibrocartilagenous disc that connects the acromion to the clavicle are brought about by previous recognized or unrecognized trauma.

Treatment

Use *rest, heat,* and *antiinflammatories.*
Steroid injection may be tried but is quite painful to perform and success rate is only fair. The method is simply to inject into the tender joint from above. See Chapter 3 for further details and precautions.

Adhesive Capsulitis (Frozen Shoulder) (*Uncommon*)

Diagnosis

The patient presents with progressive diffuse pain and global limitation of motion of the shoulder, often but not always following another recognized cause of shoulder pain.

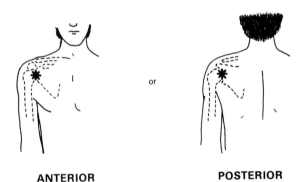

ANTERIOR   **POSTERIOR**

Fig. 6-6.   How to aspirate or inject a shoulder joint. Feel the head of the humerus and move your finger medially until it curves away from you and you feel the joint line. (You will be just lateral to the coracoid.) Alternatively, move medially on the posterior surface of the humeral head until the joint line is encountered. Prepare the site and use sterile technique throughout. After ethyl chloride spray, enter the joint perpendicularly using a 20-gauge or 18-gauge needle at least 1½ inches long, and aspirate. If steroid is to be injected, use a hemostat to hold the needle bevel and switch syringes to one containing 2 cc of steroid. (Never inject steroid if the synovial fluid is cloudy or if there is other reason to suspect infection.) Steroid can be injected the same way, without aspiration, using a 22-gauge needle. See pp. 13 and 23–27 for discussion of synovial fluid analysis and of precautions and risks *re* steroid instillation.

Pathophysiology

Fibrotic, restrictive changes in the tissues of the shoulder capsule occur secondary to disuse.

Treatment

Progressive *range of motion exercises* are essential. See Patient Handout 6-1. Include the sideways wall climb. Since these patients need close follow up to reinforce the program and check on their progress, *referral* to a *physical therapist* is wise.

Oral *antiinflammatory medications* should be used.

It may be quite helpful to use a series of three weekly intraarticular and subacromial *steroid injections*. See Figures 6-4 and 6-6.

The condition usually improves in a few months with varying degrees of residual disability.

Degenerative Arthritis (Osteoarthritis) of the GlenoHumeral Joint (*Uncommon* as a Presenting Complaint, *Common* When Many Joints Are Involved)

Diagnosis

Chronic pain is worse after use.

Found in older patients or those with previous significant trauma.

There will be variable diffuse tenderness and variable pain with all directions of motion.

Pathophysiology

See Chapter 22.

Treatment

See Chapter 22.

Technique of injection is shown in Figure 6-6.

SHOULDER TRAUMA

Always examine distal neurovascular status.

Fractures (except of the clavicle) and dislocations of the shoulder are beyond the scope of this discussion. Rotator cuff tears are mentioned only for their recognition.

Acromioclavicular Joint Sprain (Shoulder Separation, Shoulder Pointer) (*Common*)

Diagnosis

These are usually caused by a fall on the point of the shoulder.

Tenderness is found over the acromioclavicular joint (Figure 6-2).

X rays should be done with a 5-to-10 lb weight dangling in each hand and comparison made to the uninjured side. The radiograph will be normal in a first-degree sprain, will show a slight degree of displacement (not more than the width of the end of the clavicle) in a second-degree sprain, and will show a significant upward displacement of the distal end of the clavicle in a third-degree sprain.

Pathophysiology

A first-degree sprain involves an incomplete tear of the acromioclavicular ligament without joint subluxation and a second-degree injury more severe disruption of the tissues with some joint subluxation. In a third-degree sprain the coracoclavicular ligaments are torn as well.

Treatment

First-degree and mild second-degree sprains should be treated with a *sling* for one to three weeks until the pain is resolving, and then the exercises shown in Patient Handout 6-1 should be begun to restore strength and maintain range of motion.

Severe second-degree sprains and all third-degree sprains should be *referred* to an orthopedist for management.

Clavicle Fractures (*Common*)

Diagnosis

These are usually due to a fall on the outstretched hand or on the point of the shoulder.

Treatment

Fractures at the medial or lateral ends of the clavicle should be referred to an orthopedist for management.

Other fractures may be treated with a *figure-of-eight strap* applied tightly until abduction is pain-free (usually about four weeks in an adult). A sling should be used as well for the first week or so. Referral will be necessary if satisfactory reduction cannot be maintained.

Rotator Cuff Tears (*Uncommon*)

Diagnosis

These occur following acute trauma (a fall on the abducted arm) or acute strain (e.g., throwing).

Degeneration of the cuff tendons predisposes, and therefore in older patients the responsible trauma may be much milder.

On examination, tenderness is found over the rotator cuff and there is weakness or loss of active elevation past a certain point. The latter is sometimes difficult to differentiate from unwillingness to elevate the arm secondary to pain from tendonitis.

Referral for arthrography may be necessary to make the diagnosis.

Management

The patient should be *referred* to an orthopedist for consideration of surgery, except in cases of partial degenerative tears where the disability from the tear does not warrant such consideration.

Patient Handout 6-1

Shoulder Mobilization Exercises

These exercises are very important to prevent your shoulder from stiffening up, or to restore a full range of motion if some restriction has already occurred. Do them twice a day.

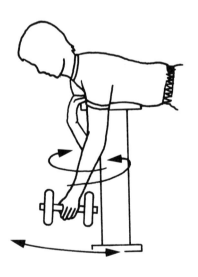 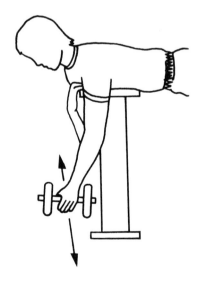

PENDULUM EXERCISE

Do this with a weight in your hand. Move your arm forward and back further and further with each swing until it's swinging as far as it will go; at that point, swing ten more times. Then do the same thing side-to-side. Then make circles in one direction, swinging wider and wider; then the same in the other direction.

WALL CLIMBING

Face the wall with your arm outstretched and your fingertips just touching; now without leaning your body, walk your fingertips up the wall as far as they will go. Repeat five times.

☐ Do the same thing with your side toward the wall and your arm elevated at your side. Do this one only if your doctor has checked the box.

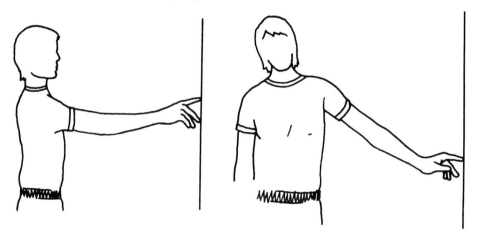

7

The Elbow

Elbow problems in *toddlers* and *teenagers* are discussed in Chapter 19. Further details on therapeutic suggestions made in this chapter can be found in Chapter 3.

For the anatomy of the elbow see Figure 7-1.

ELBOW PAIN

History

Ask about the *duration* and *location* of the pain. (Occasionally a diffuse ache may be *referred pain* from cervical or intrathoracic disease: it is wise to keep this in mind).

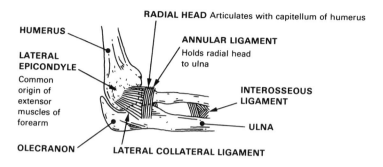

RADIAL HEAD Articulates with capitellum of humerus

HUMERUS

LATERAL EPICONDYLE

Common origin of extensor muscles of forearm

ANNULAR LIGAMENT
Holds radial head to ulna

INTEROSSEOUS LIGAMENT

ULNA

OLECRANON

LATERAL COLLATERAL LIGAMENT

LATERAL VIEW

ANTERIOR VIEW OF DISTAL HUMERUS

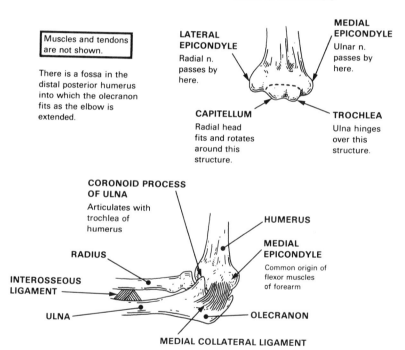

Muscles and tendons are not shown.

There is a fossa in the distal posterior humerus into which the olecranon fits as the elbow is extended.

LATERAL EPICONDYLE
Radial n. passes by here.

MEDIAL EPICONDYLE
Ulnar n. passes by here.

CAPITELLUM
Radial head fits and rotates around this structure.

TROCHLEA
Ulna hinges over this structure.

CORONOID PROCESS OF ULNA
Articulates with trochlea of humerus

RADIUS

INTEROSSEOUS LIGAMENT

ULNA

HUMERUS

MEDIAL EPICONDYLE
Common origin of flexor muscles of forearm

OLECRANON

MEDIAL COLLATERAL LIGAMENT

MEDIAL VIEW

Fig. 7-1. Anatomy of the elbow.

What are *precipitating* and *relieving factors*?
Has there been any specific *trauma, strain,* or *overuse*?
Inquire about any *previous elbow problems,* diagnoses, and treatments.
Are any *other joints* involved, or is there a history of arthritis or gout?
Does the joint ever *lock*? (True locking indicates a loose body in the joint from previous injury, osteochondritis dissecans, or severe degenerative changes. Take an X ray and, if symptoms are significant enough to warrant treatment, refer the patient to an orthopedist.)

Examination

Localize *tenderness* (Figure 7-2).
Look for *swelling* (Figure 7-3).
Of course, any *redness* or *warmth* implies an acute inflammatory process.
Test *range of motion*: pronation/supination (palm up/palm down; normal is 90° each way) as well as flexion/extension (normal is 0 to 160°). (*Note:* A flexion contracture, i.e., lack of full extension, is a nonspecific sign of long-term elbow disease and disuse.) Painful range of motion in the absence of acute trauma indicates an intra-articular process, most likely degenerative arthritis in chronic cases, crystals or infection in acute ones.
Check *distal pulses, strength,* and *sensation.*

Lateral Epicondylitis (Tennis Elbow) (*Very Common*)

Diagnosis

Tenderness is at or near the lateral epicondyle (Figure 7-2).
Confirm by pain referred there on resisted wrist extension and/or supination.
Elbow range of motion is essentially full and painfree. (Sometimes with severe epicondylitis, there may be some pain at extreme flexion or extension.)

Pathophysiology

Inflammation (and possibly some microtears) occurs where the common extensor tendon of the forearm attaches, precipitated by acute or chronic use of the extensor and supinator muscles of the forearm.
Common with various occupations as well as with racket sports.

Treatment

Have the patient *avoid* the responsible activity as much as possible.
A *wrist splint* is especially useful where the extensor muscle overuse is with wrist movement (as opposed to isometric): in a house painter, for example.

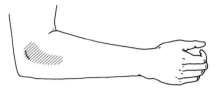

LATERAL EPICONDYLITIS
There will be pain here as you resist the
patient's effort to extend (dorsiflex) his
wrist and/or supinate his forearm.

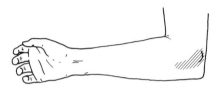

MEDIAL EPICONDYLITIS
There will be pain here as you resist the
patient's effort to flex his wrist and/or
pronate his forearm.

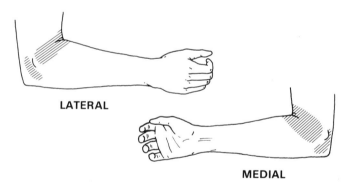

LATERAL

MEDIAL

An Intraarticular Process
Consider DEGENERATIVE ARTHRITIS in an older
patient or one with previous trauma; if there
is redness, swelling, or warmth, the joint must
be tapped to rule out infection (See Fig. 7-3).

Fig. 7-2. Sites of tenderness: the elbow.

A *tennis elbow band* (a *non*elastic strip applied circumferentially around the
proximal forearm) theoretically prevents excessive muscle contraction by con-
striction, and is more useful in isometric overuse of the muscle.

In those who play racket sports, advise a lighter racket with smaller grip and

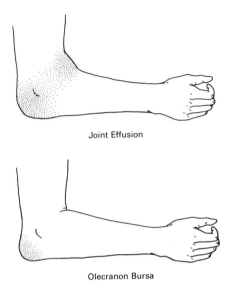

Joint Effusion

Olecranon Bursa

Fig. 7-3. Elbow swelling.

less string tension) using a two-handed back-hand, and not applying excessive backspin.

Note that immobilization of the elbow is irrelevant and useless.

Antiinflammatory medications may be helpful.

Steroid injection may be offered. Success rate is only fair; see Figure 7-4.

Cold is helpful in acute cases and pre- and post-exercise; *heat* may give more relief in chronic cases.

The program in Patient Handout 7-1 is useful for rehabilitation and to prevent recurrence once the acute episode has calmed. Some feel it should be started

Fig. 7-4.   Injection for epicondylitis. Prep the area and spray with ethyl chloride. Use a 23-gauge needle to inject a cc of steroid and a cc of local anesthetic in the manner shown. Injections may be repeated weekly or biweekly to a total of three or so. See pp. 23–27 for further details and precautions.

when pain is still present in long-term cases, to try to form a painfree scar at the muscle origin.

Referral to a physical therapist for *transverse friction massage* can sometimes be helpful, although it is painful to the patient to perform.

Surgery should be a last resort only, since most cases of epicondylitis improve in a few months. The usual procedure is to remove the common extensor tendon from its origin and reattach it further distally.

Note that some patients with what seems to be refractory epicondylitis may actually have the RADIAL TUNNEL SYNDROME, which is entrapment of the posterior interosseous nerve (a branch of the radial) as it passes under the supinator muscle. Tenderness will be localized over muscle, an inch or so distal, and a bit toward the antecubital area, from the epicondyle. Pain will be referred to that point with resisted middle finger extension and forearm supination (done with the elbow straight). Surgical release has a very good success rate.

Medial Epicondylitis (*Common*)

Diagnosis

Tenderness is at the medial epicondyle (Figure 7-2).
Confirm by pain referred there on resisted wrist flexion and/or pronation.
There is usually full and pain-free range of motion of the elbow.

Pathophysiology

Analogous to lateral epicondylitis, but with overuse of the flexor and pronator muscles of the forearm.
Common with throwing and golfing.

Treatment

Identical to lateral epicondylitis.

Degenerative Arthritis (Osteoarthritis) of the Elbow
(*Uncommon* as a Presenting Complaint, *Common* When
Many Joints Involved)

Diagnosis

Chronic pain is increased after use.
Found in older patients or those with previous significant trauma.

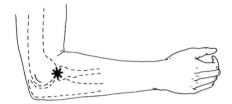

Fig. 7-5. How to aspirate or inject an elbow joint: (a) Feel the tip of the lateral epicondyle with the elbow at 90° of flexion. Now move just distal, to it (toward the hand) and feel the joint line between the humerus and the radial head. (b) Prepare the site and use sterile technique throughout. After ethyl chloride spray, enter the joint using a 22- or 20-gauge needle and aspirate. (c) If steroid is to be injected, use a hemostat to hold the needle bevel and switch syringes to one containing 2 to 4 cc of steroid. (Never inject steroid if the synovial fluid is cloudy or if there is other reason to suspect infection.) Steroid can be injected the same way, without aspiration, using a 22-gauge needle. Be sure to see pp. 13 and 23–27 for discussion of synovial fluid analysis and of precautions and risks *re* steroid instillation.

There is variable diffuse tenderness and variable pain with range of motion. There may be slight swelling in an acute exacerbation, but if it is red or warm it must be tapped to rule out an infection.

The X ray may be normal early (when changes are limited to joint cartilage), but it is characteristic later. (But remember that the presence of degenerative changes on X ray does not prove that any particular patient's symptoms are due to those changes.)

Pathophysiology

See Chapter 22.

Treatment

See Chapter 22.
Technique of injection is shown in Figure 7-5.

ELBOW SWELLING

Evaluation

Swelling of the olecranon bursa must be differentiated from swelling of the joint itself, which implies an intraarticular process. See Figure 7-3.

Olecranon Bursitis (*Common*)

Diagnosis

A painless or slightly painful swelling is localized over the olecranon (Figure 7-3).
If there is redness, warmth, significant tenderness, or a nearby break in the skin, it must be tapped to rule out infection.

Pathophysiology

Fluid accumulates in the inflamed bursa, usually secondary to leaning on the elbow (especially on a hard surface).
Occasionally infection, and rarely gout, may be responsible.

Treatment

The patient must *avoid* leaning on the elbow; a *pad* may be used to prevent recurrence if leaning is unavoidable.
Some feel that immobilizing the elbow in a *sling* for a short time helps by avoiding rubbing the bursa with elbow flexion and extension.
Heat may help.
Drainage may be offered; the fluid usually reaccumulates very quickly unless some *steroid* is instilled after drainage, but even this often does not work. Technique is simply to enter the center of the swelling with a #18 needle (with a large syringe) after sterile preparation and ethyl chloride spray; never inject steroid if the fluid drained appears cloudy or purulent. Remember, the bursa must be aspirated if the signs of infection described above are present.

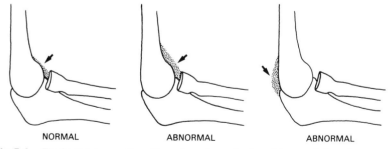

NORMAL   ABNORMAL   ABNORMAL

Fig. 7-6.   The fat pad sign on elbow X ray. A small anterior fat pad is normal. But a large one, or any visible posterior fat pad, is indicative of an intraarticular fracture even if the fracture is not visualized. (And they can be notoriously hard to see.)

Surgery may be done to remove a recurrently or chronically inflamed bursa (which may have a lumpy feeling to it, by the way).

ELBOW TRAUMA

Always examine distal neurovascular status.

Note: If there is significant pain or swelling in the elbow after direct or indirect trauma (such as a fall on the outstretched hand), or a positive fat pad sign (Figure 7-6) on X ray, suspect a fracture even if not visualized, and refer the patient to an orthopedist.

Elbow Sprains (*Uncommon*)

Diagnosis

Usually caused by a hyperextension injury.

There is diffuse soft tissue swelling and tenderness.

No fracture or fat pad sign (Figure 7-6) is seen on X ray.

Treatment

If severe, referral to an orthopedist is best.

If mild with a negative X ray, treat with *ice* intermittently for the first 48–72 hours, with *analgesia* and *rest* (including use of a sling).

Early active range of motion should be instituted to prevent a flexion contracture. When to begin depends on clinical progress.

Patient Handout 7-1

Rehabilitation Exercises for Epicondylitis

Put your arm on a table with your wrist hanging over the edge, holding something in your hand that weighs a couple of pounds. Now *slowly* straighten your wrist and bend it upwards; then slowly turn it palm up. Now *slowly* turn it back palm down and lower it. Do 15 repetitions twice a day. Increase the weight by a couple of pounds every few days, until you are up to 10 or 15 pounds.

8

The Wrist

For the anatomy of the wrist see Figure 8-1.

WRIST PAIN

History

Ask about the *duration* and *location* of pain.
What are *precipitating* and *relieving factors*?
Has there been any specific *trauma, strain,* or *overuse*?
Inquire about any *previous wrist problems,* diagnoses, or treatments.
Are any *other joints* involved? Is there a history of arthritis or gout?

1. NAVICULAR (SCAPHOID)	5. TRAPEZIUM
Most often fractured	(GREATER MULTANGULAR)
2. LUNATE	6. TRAPEZOID
Second most commonly fractured	(LESSER MULTANGULAR)
3. TRIQUETRUM	7. CAPITATE
4. PISIFORM	8. HAMATE
Volar to other bones	

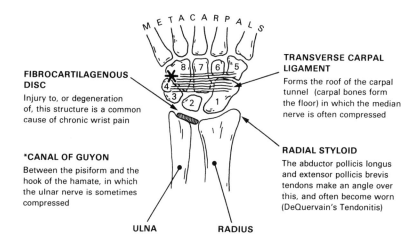

FIBROCARTILAGENOUS DISC
Injury to, or degeneration of, this structure is a common cause of chronic wrist pain

TRANSVERSE CARPAL LIGAMENT
Forms the roof of the carpal tunnel (carpal bones form the floor) in which the median nerve is often compressed

***CANAL OF GUYON**
Between the pisiform and the hook of the hamate, in which the ulnar nerve is sometimes compressed

RADIAL STYLOID
The abductor pollicis longus and extensor pollicis brevis tendons make an angle over this, and often become worn (DeQuervain's Tendonitis)

ULNA   **RADIUS**

Many tendons pass over the wrist. Those on the dorsal surface are covered by a synovial sheath.
There are also a multitude of ligaments (not shown) interconnecting the bones.

Fig. 8-1. Anatomy of the wrist.

If pain radiates into the hand or up the arm or is associated with numbness or paresthesias, see Chapter 5.

Examination

Localize *tenderness* (See Figure 8-2).

Any *generalized* swelling in the absence of trauma implies an intraarticular process, which if acute *must be tapped* to rule out infection.

Any *redness* or *warmth* of course also implies an acute inflammatory process.

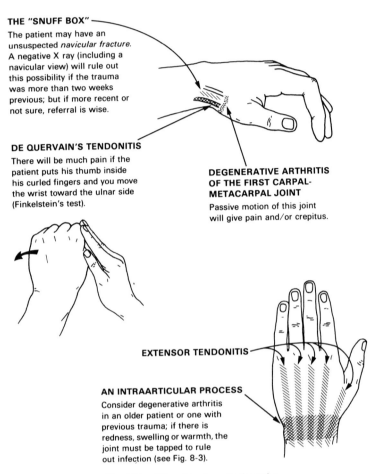

THE "SNUFF BOX"
The patient may have an unsuspected *navicular fracture.* A negative X ray (including a navicular view) will rule out this possibility if the trauma was more than two weeks previous; but if more recent or not sure, referral is wise.

DE QUERVAIN'S TENDONITIS
There will be much pain if the patient puts his thumb inside his curled fingers and you move the wrist toward the ulnar side (Finkelstein's test).

DEGENERATIVE ARTHRITIS OF THE FIRST CARPAL-METACARPAL JOINT
Passive motion of this joint will give pain and/or crepitus.

EXTENSOR TENDONITIS

AN INTRAARTICULAR PROCESS
Consider degenerative arthritis in an older patient or one with previous trauma; if there is redness, swelling or warmth, the joint must be tapped to rule out infection (see Fig. 8-3).

Fig. 8-2.   Sites of tenderness: the wrist.

Test *range of motion* (normal is 80° flexion, i.e., turned toward the palm, 70° extension, 30° ulnar deviation, and 20° radial deviation).

Check *strength, sensation* and *distal capillary filling.*

DeQuervain's Tendonitis (*Very Common*)

Diagnosis

Tenderness is found over the radial styloid and/or the tendons that pass over it (Figure 8-2).

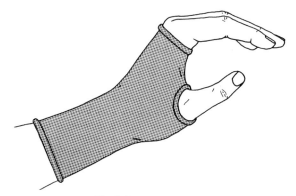

Fig. 8-3.   Wrist splint.

Confirm by eliciting much pain with Finkelstein's test (Figure 8-2), and often also with pain on resisting the patient's attempt to extend his thumb.

Pathophysiology

Inflammation of the abductor pollicis longus and extensor pollicis brevis tendons (which move in the same sheath) results from friction where they form a sharp angle over the radial styloid.

Common with knitting, the use of a wrench, holding a baby, and various other activities.

Treatment

Advise the patient to *avoid* the responsible activity as much as possible. Use a *wrist splint* (Figure 8-3). (This is somewhat helpful but incompletely so,

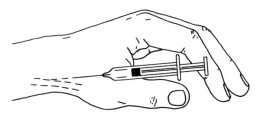

Fig. 8-4.   Injection For DeQuervain's tendonitis. Prep the area and spray with ethyl chloride. Enter the tendon sheath with a 25-gauge needle as shown. (One way to do this is to enter the tendon and slowly pull back until the needle pops out.) Inject half a cc each of steroid and of local anesthetic without epinephrine. If you are in the sheath you will see the fluid distend it in a linear pattern. Never inject into the tendon itself. See pp. 23–27 for further details and precautions.

since the thumb is not immobilized. Also, be sure part of the splint is not exerting pressure over the sore area.)

Instead, a short arm *cast* with a thumb extension may be resorted to in order to achieve adequate immobilization.

Antiinflammatory medications and *heat* are helpful.

Steroid injection may be offered. Success rate is quite good. See Figure 8-4.

In prolonged or recurrent cases when conservative measures and injection have failed, referral to an orthopedist for *surgery* (in which the tendon sheath is split longitudinally) should be considered.

Wrist Tendonitis (*Common*)

Diagnosis

Linear tenderness is found over a specific tendon.

Range of motion is painful only when stressing the involved tendon.

There is *no* redness or warmth. (Infectious tenosynovitis is not uncommon here; if suspected, immediate referral to an orthopedist is mandatory.)

Pathophysiology

Tendon inflammation occurs from minor strain or overuse.

Treatment

Use a wrist splint (Figure 8-3).

Antiinflammatory medications and *heat* are helpful.

Steroid injection around (not into) the involved tendon may be offered in long-term cases. See pp. 23–27 for details and precautions.

Degenerative Arthritis of the Carpometacarpal Joint of the Thumb (*Common*)

Diagnosis

Tenderness is found around that joint (Figure 8-2).

There is pain on motion of that joint, often with crepitus.

Pathophysiology

Accelerated wear and tear occurs here from carrying objects in a pinch between thumb and fingers.

Common in mail carriers.

Treatment

The patient should *avoid* the responsible activity as much as possible.
Steroid injection is usually helpful. See Figure 8-5.
If the patient declines injection, oral *antiinflammatory medication* may be used.
Heat may help.

Degenerative Arthritis of the Wrist (Diffuse) (*Uncommon* as a Presenting Complaint, *Common* When Many Joints Involved)

Diagnosis

Chronic pain is worse after use.
Found in older patients or those with previous trauma.
There is variable diffuse tenderness and variable pain with range of motion.
There may be slight swelling during an acute exacerbation, but if it is red or warm it must be tapped to rule out infection.

Pathophysiology

See Chapter 22.

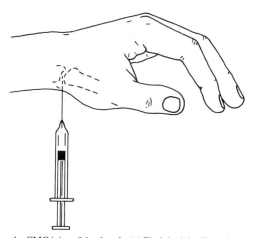

Fig. 8-5. Injecting the CMC joint of the thumb. (a) Find the joint line with your thumb by putting your fingers on the thenar eminence and moving the metacarpal. (b) Prep. the area and spray with ethyl chloride. (c) Inject into the dorsal radial aspect with a 25-gauge needle ½ cc each of steroid and local anesthetic without epinephrine. See p. 23 for further details and precautions.

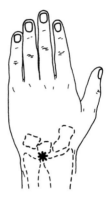

Fig. 8-6. How to aspirate or inject a wrist joint. (a) Feel the radius and ulna on the dorsal aspect of the wrist; there is a palpable depression between their distal ends. (b) Prepare the site and spray with ethyl chloride. Use sterile technique throughout. Enter the joint perpendicularly using a 22-gauge or 20-gauge needle, and aspirate. (c) If steroid is to be injected, use a hemostat to hold the needle bevel and switch syringes to one containing 2 cc of steroid. (Never inject steroid if the synovial fluid is cloudy or if there is other reason to suspect infection.) Steroid can be injected the same way, without aspiration, using a 22-gauge needle. Be sure to see pp. 13 and 23 for discussion of synovial fluid analysis and of precautions and risks *re* steroid instillation.

Treatment

See Chapter 22.
Technique of injection is shown in Figure 8-6.

WRIST LUMPS

Ganglia (*Very Common*)

Diagnosis

The lumps can occur at any location around the wrist, and may come and go. Increase in size is often related to joint usage.

They are sometimes painful, and occasionally can cause symptoms of nerve compression.

On examination, a smooth, rounded cystic lump is found, which is nonpulsatile and usually $\frac{1}{2}$ to 2 cm in size.

If the lump is *progressively growing* or *not obviously cystic*, the patient should be *referred* for *biopsy* to rule out a possibly neoplastic lesion, though this would be extremely rare.

Pathophysiology

The cyst arises from a tendon sheath or from the joint synovium.

Treatment

If it is not progressively growing by history and is obviously cystic on examination, and if it does not bother the patient, no treatment is necessary.

A wrist splint can be symptomatically helpful if the patient is not interested in more definitive treatment.

The ganglion may be *aspirated*; see Figure 8-7. Recurrence is common.

If this method is unsuccessful or if the ganglion recurs, the patient may be offered referral to an orthopedist for *surgical removal*. Excision can be somewhat complex, especially in those ganglia that arise from the joint synovium, due to the necessity for tracing and removing multiple long roots of the ganglion. Failure to do this completely accounts for some cases of postoperative recurrence.

Fig. 8-7. Aspirating a ganglion. (a) Flex the wrist over a pillow or roll as shown. Make sure the mass is not pulsatile. (b) After prep and ethyl chloride spray, enter the ganglion with an 18-gauge needle and aspirate. (c) Whether or not the thick gelatinous contents of the ganglion are obtained, use finger pressure to flatten the cyst into disappearance. Multiple punctures and/or the instillation of a small amount of steroid may decrease the recurrence rate. If the cyst cannot be flattened, it may not be a ganglion, and must be referred for excisional biopsy.

WRIST TRAUMA

Evaluation

Those who make the often-heard statement ''there is no such thing as an acute wrist sprain'' may be carrying things a bit far, but carpal bone *fractures*, especially of the navicular (scaphoid) bone, are notorious for *not showing up on early X rays* and for leading to serious consequences if not adequately treated. The navicular in particular is prone to avascular necrosis after a fracture, due to its tenuous blood supply.

Furthermore, certain specific ligamentous injuries of the wrist need early specific treatment if disability is to be minimized.

Therefore any acute wrist injury with any swelling or significant tenderness [or even mild tenderness if located over the navicular (Figure 8-2)] should be referred to an orthopedist.

Always check distal neurovascular status.

Fractures and dislocations are beyond the scope of this book.

Wrist Sprain (*Common*)

Diagnosis

See the caution above.
Usually due to a hyperextension injury.

Treatment

Use a *wrist splint, analgesics,* and for the first 48 hours, *cold* and *elevation.*

9

The Hand

Further details on therapeutic suggestions made in this chapter can be found in Chapter 3.

For the anatomy of the hand see Figure 9-1.

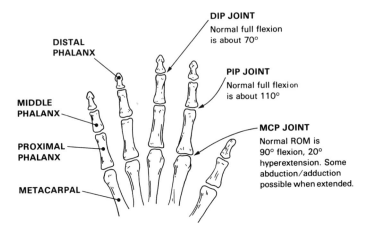

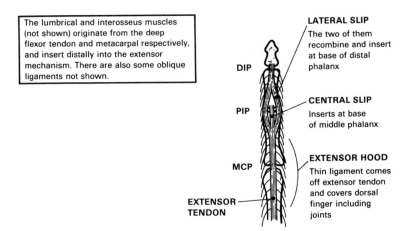

**DORSAL VIEW OF
FINGER SHOWING
EXTENSOR MECHANISM**

Fig. 9-1.   Anatomy of the hand.

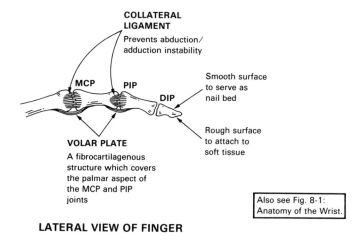

COLLATERAL LIGAMENT
Prevents abduction/adduction instability

MCP PIP

DIP

Smooth surface to serve as nail bed

VOLAR PLATE
A fibrocartilagenous structure which covers the palmar aspect of the MCP and PIP joints

Rough surface to attach to soft tissue

Also see Fig. 8-1: Anatomy of the Wrist.

LATERAL VIEW OF FINGER

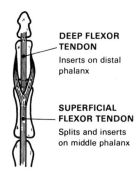

Note: The anatomy of the hand and digits is extremely complex; many fascial layers, septae, ligaments, tendons, and intrinsic muscles are not shown here. Nor, obviously, are nerves and vascular structures.

DEEP FLEXOR TENDON
Inserts on distal phalanx

SUPERFICIAL FLEXOR TENDON
Splits and inserts on middle phalanx

PALMAR VIEW OF FINGER SHOWING FLEXOR TENDONS

Fig. 9-1. Continued.

DIFFUSE NONLOCALIZED HAND PAIN

Evaluation

A neurovascular examination is essential.

Diffuse hand aching, without localized tenderness on examination, especially if worst at night, is suggestive of CARPAL TUNNEL SYNDROME, which is discussed in Chapter 5. Other radiated pain (from the neck or thoracic outlet) should also be considered. See Chapter 5.

Diffuse hand or finger pain which comes on with exposure to cold and is relieved with rewarming suggests *Raynaud's Phenomenon*. The reader is referred to a textbook of rheumatology or internal medicine for further discussion of this condition.

ATRAUMATIC PAIN IN THE JOINTS OF THE HAND AND FINGERS

Evaluation

Acute atraumatic monoarthritis in the hand has the same significance as elsewhere, with the leading possibilities being infection or crystalline disease; the joint must be tapped, especially to rule out the former possibility.

Three types of arthritis commonly affect the joints of the hand and fingers: OSTEOARTHRITIS, RHEUMATOID ARTHRITIS, and PSORIATIC ARTHRITIS.

OSTEOARTHRITIS more commonly affects the DIP and PIP joints; morning stiffness lasts only a few minutes; examination will commonly reveal hypertrophy at the DIP joints (Heberden's nodes) and less commonly, the PIP joints (Bouchard's nodes). Redness and warmth are uncommon; X ray will show joint space narrowing and hypertrophic changes of bone; blood tests such as the ESR and rheumatoid factor will be normal. See Chapter 22.

One clinical type of PSORIATIC ARTHRITIS involves primarily the DIP joints. Involvement is usually asymmetric; the joint disease may become manifest before any skin lesions are apparent, but usually involved fingers will be found on close inspection to have pitting of their nails also. X ray will show some bony erosion, especially in the tufts of the distal phalanges; the sedimentation rate may be elevated, while the rheumatoid factor will be negative. See Chapter 23.

RHEUMATOID ARTHRITIS in the hands involves mainly the PIP and MCP joints. Morning stiffness will be a significant complaint; systemic symptoms can be present. Inflammation will usually be evident in acute cases, and synovial thickening will be palpable in chronic ones. Various deformities can be found in more advanced cases. The X ray may be normal early, but will eventually begin to show erosions of bone at the joint margins. The ESR will be elevated, and the rheumatoid factor will usually be positive. See Chapter 23.

SNAPPING FINGER

Trigger Finger (or Trigger Thumb) (*Common*)

Diagnosis

The patient will complain, and examination will confirm, that when the involved digit is flexed there is resistance to reextension in mid-arc which sometimes can only be overcome by extending the digit passively with the other hand. Reextension is accompanied by a palpable and sometimes audible pop.

There will often be tenderness (and sometimes a palpable lump as well) at the base of the flexor tendon sheath.

Pathophysiology

A thickening develops in the flexor tendon of the digit, and resistance to entrance into the base of the flexor tendon sheath is encountered as the digit is extended.

The thickening is most usually a consequence of overuse but can be related to rheumatic disease.

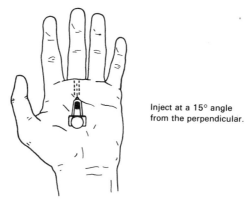

Inject at a 15° angle
from the perpendicular.

Fig. 9-2. Injection for trigger finger. Find the tender area at the base of the flexor tendon sheath (just distal to the distal palmar crease for the fingers, at the level of the MCP crease for the thumb). After prep and ethyl chloride spray, enter the tendon as shown with a 25-gauge needle and slowly withdraw until you just feel it come out of the tendon. Now inject about a quarter of a cc each of steroid and local anesthetic without epinephrine (mixed in the same syringe, of course). Never inject against resistance. See Chapter 3 for further details and precautions.

Treatment

Success rate of *steroid injection* is good if done proeprly. See Figure 9-2. It may have to be repeated at intervals.

Referral for *surgery* (in which the tendon sheath is split) is the only other treatment option available.

FRACTURES

Types

Fractures of the tuft of the distal phalanx are discussed below.

Avulsion fractures at the base of the phalanges are treated as the corresponding sprain (Table 9-I), unless the fragment involves more than about one-fourth of the joint surface, in which case the patient should be referred to an orthopedist.

The following fractures require reduction or at least the application of plaster; the reader is advised to refer to a textbook of orthopedics or to refer the patient for proper treatment.

1. Fractures of the metacarpals.
2. Transverse, oblique, or spiral fractures of the proximal or middle phalanges.
3. Transverse fractures of the proximal portion of the distal phalanx.

Tuft Fracture of the Distal Phalanx (*Common*)

Diagnosis

These are usually sustained by crushing-type injuries.

If there is a transverse fracture of the proximal end of the distal phalanx, refer the patient for treatment.

Table 9-I   Avulsion Fractures

Avulsion fracture	Corresponding sprain
Corner of base of middle phalanx	Collateral ligament
Volar base of middle phalanx	Volar plate injury
Dorsal base of middle phalanx	Central extensor slip tear
Volar base of distal phalanx	Flexor profundus tear (REFER)
Dorsal base of distal phalanx	Mallet finger

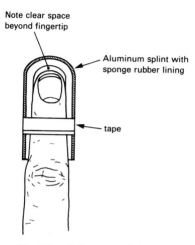

Note clear space beyond fingertip

Aluminum splint with sponge rubber lining

tape

Fig. 9-3. Splint for a tuft fracture.

Management

Tuft fractures, even when severely comminuted, can be treated simply by preventing further trauma with a *splint* as shown in Figure 9-3.

DISLOCATIONS

Dislocations of the MCP joint and their reduction can lead to complications; their management is best left to the orthopedist.

PIP Joint (*Common*)

Diagnosis

This is usually due to a hyperextension injury.
An X ray should be done to rule out an associated fracture.

Management

Reduction can be done as shown in Figure 9-4. Anesthesia is usually unnecessary.
After reduction, check to be sure that there is not hyperextensibility or volar midline tenderness, or conversely, lack of full active extension of the joint. If any

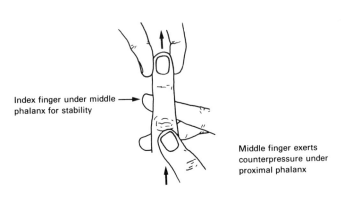

Other hand pulls
axially gently

Index finger under middle
phalanx for stability

Middle finger exerts
counterpressure under
proximal phalanx

Thumb pushes distally
and downward

Fig. 9-4. Reduction of dislocation of the PIP joint.

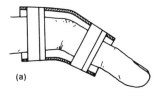

(a)

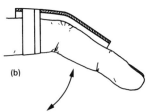

(b)

Fig. 9-5. (a) 30° to 40° flexion splint for the PIP joint. (b) 30° Extension block splint.

of these are found, the joint should be treated as for a VOLAR PLATE INJURY or CENTRAL EXTENSOR SLIP TEAR, respectively.

If the findings in the previous paragraph are not present, *splint* the joint at about 40° of flexion (Figure 9-5) for about ten days.

Ice packs will be helpful for the first several days.

SPRAINS OF THE MCP JOINT OF A FINGER

Sprains of these joints usually involve at least some tearing of the extensor hood. This will manifest itself by the patient's inability to extend the joint actively from a fully flexed position, often with visible or palpable slippage of the extensor tendon off the dorsum of the joint. All significant injuries to this joint are best managed by the orthopedist.

SPRAINS OF THE MCP JOINT OF THE THUMB

These are almost always injuries to the ulnar collateral ligament of the joint.

Sprain of the Ulnar Collateral Ligament of the MCP Joint of the Thumb (Gamekeeper's Thumb) (*Common*)

Diagnosis

The patient usually gives a history of falling onto the inside of the thumb with the thumb held away from the fingers.

Especially common in skiers and football players.

Management

These injuries can lead to significant disability if inadequately treated, since stability of pinch can be lost. Therefore, if there is any significant tenderness or any suggestion of instability (see p. 111), or an avulsion fraction on X ray, the patient should be *referred* to an orthopedist for management (which consists of several weeks in a thumb spica cast for stable injuries, or surgical repair for unstable ones).

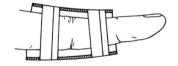

Fig. 9-6. Extension splint for the PIP joint.

SPRAINS OF THE PIP JOINT

Evaluation

If there is tenderness over the volar base of the middle phalanx, or if the joint can be passively *hyperextended,* or if there is an avulsion fracture at the volar base, the patient has a VOLAR PLATE INJURY.

If the patient cannot *fully* actively extend the PIP joint, he must be considered to have a CENTRAL EXTENSOR SLIP TEAR. The classic "boutonniere" deformity does not develop until later.

If the above findings are absent, consider the patient to have a COLLATERAL LIGAMENT SPRAIN.

Volar Plate Injury (*Uncommon*)

Diagnosis

The history will usually be of a hyperextension injury.

There will be tenderness over the volar PIP joint.

The PIP joint can be passively hyperextended (when compared to the other hand).

Management

If there is an avulsion chip which is at all large (more than about 15% of the joint) or displaced (the so-called WILSON'S FRACTURE), the patient should be referred to the orthopedist. Internal fixation may be necessary.

Otherwise the injury should be splinted in about 30° of flexion for ten days, and then left in a 30° extension block splint for an additional ten days (Figure 9-5). During the latter period active motion in the splint is encouraged. Passive and active range of motion must be faithfully performed for several weeks after the splint is removed to prevent subsequent stiffness.

Central Extensor Slip Tear (*Uncommon*)

Diagnosis

Usually a history of sudden resisted flexion will be given.

The patient is unable actively to extend completely the joint.

Management

The joint should be splinted in hyperextension (p. 109) for three weeks, and then active motion begun. Inadequate or delayed treatment will lead to the development of the classic "boutonniere" deformity, which must then be surgically corrected.

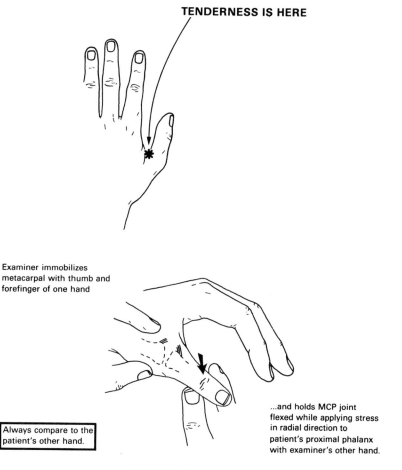

TENDERNESS IS HERE

Examiner immobilizes metacarpal with thumb and forefinger of one hand

...and holds MCP joint flexed while applying stress in radial direction to patient's proximal phalanx with examiner's other hand.

Always compare to the patient's other hand.

TESTING STABILITY OF ULNAR COLLATERAL LIGAMENT OF THUMB

Fig. 9-7. Gamekeeper's thumb.

Collateral Ligament Sprain (*Very Common*)

Diagnosis

Usually the mechanism is axial trauma (the "jammed finger").
The findings of VOLAR PLATE INJURY or CENTRAL EXTENSOR SLIP
TEAR are absent.
Tenderness is maximal on the side of the joint.
Check for lateral stability with the joint in 15° of flexion.

Management

If there is an avulsion fracture of more than about 15% of the joint surface or
gross instability, *refer* the patient to an orthopedist.
Otherwise, *splint* the joint in about 40° *of flexion* for 10 days, and then begin
active motion.
Ice packs are helpful acutely.

SPRAINS OF THE DIP JOINT

Evaluation

See if the patient is able to flex the joint actively. If not, he probably has an
avulsion of the flexor profundus tendon and should be sent to the orthopedist for
repair.
Check to see if there is a loss of full active extension at the joint. If so, see
MALLET FINGER, below.
If X ray is normal with neither of the above findings, no treatment is necessary.

Mallet Finger (Baseball Finger, Dropped Finger)
(*Common*)

Diagnosis

Usually the history is one of a blow on the dorsal part of the distal phalanx
forcing sudden flexion.
The patient will be unable to extend the DIP joint fully.

The DIP joint must be held in full
extension or slight hyperextension.

Fig. 9-8. Mallet finger splint. If in hyperextension, be careful to avoid blanching of the skin over the dorsal joint, which could lead to sloughing. Various commercial splints are available for the treatment of this injury.

Management

If there is a large (greater than about one-fourth of the joint surface) avulsion fragment, *refer* the patient to the orthopedist.

Otherwise, *splint* the joint in *hyperextension* for six weeks. The trick is to impress upon the patient the need never to allow the fingertip to fall into flexion even for a moment until treatment is complete. See Figure 9-8. If treatment is delayed, the chance for successful nonsurgical healing is decreased.

CONTRACTURES

Obviously contractures of the various joints can occur with previous trauma or with denervation. DUPUYTREN'S CONTRACTURE is a thickening of the

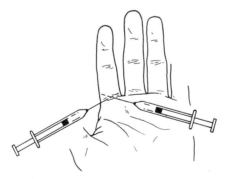

Fig. 9-9. Anesthetizing a finger. After prep and ethyl chloride spray, inject about a cc of local anesthetic *without* epinephrine toward the medial and lateral sides of the finger from a point in the midline of the proximal crease. Use a 25-gauge needle. It usually takes about ten minutes for the anesthetic to take effect.

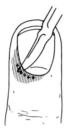

xxx shows where
incision may be
made if lifting
cuticle doesn't lead
to sufficient drainage.

Fig. 9-10. Draining a paronychia. Perform a digital block as shown in Figure 9-9. After prepping the area, use a scalpel blade to lift the cuticle where infection exists. Sometimes it will be necessary to make a skin incision as well. Place a small strip of iodoform gauze under the cuticle and apply a dressing. Advise the patient to soak the finger in warm water for about thirty minutes every few hours. Also prescribe a course of antistaphylococcal antibiotics.

palmar fascia that eventually leads to flexion contracture of the fourth and/or fifth digits. It is often associated with cirrhosis of the liver, but can be idiopathic. Treatment is by surgical release.

HAND LUMPS

Evaluation

A tiny hard lump in the midline of the volar base of a digit is usually a small *sesamoid* bone in the flexor tendon sheath.

An extremely tender nodule following a laceration is probably a *neuroma,* which must be treated surgically.

GANGLIA are not rare on the hand; see discussion of this topic in Chapter 8.

Other lumps (which are not simply benign-appearing exostoses on X ray) should be referred for biopsy.

HAND INFECTIONS

A method for draining a PARONYCHIA (subcuticular infection) is described in Figure 9-10. Infections of any of the various other spaces of the hand are potentially disastrous and should be referred STAT to a hand surgeon.

10

The Chest

CHEST PAIN

Note: Although musculoskeletal chest pain is discussed below, far more important is ruling out much more serious and sometimes life-threatening pathology, including conditions of the cardiac, respiratory, and gastrointestinal systems. The differential diagnosis of chest pain is beyond the scope of this book; suffice it to say that no complaint of chest pain should ever be taken to be musculoskeletal in origin until other etiologies have been absolutely excluded by history, examination, and whatever tests are deemed appropriate.

Musculoskeletal Chest Pain (Costochondritis, Pulled Chest Muscle) (*Very Common*)

Diagnosis

The pain may be intermittent or constant, the duration from hours to years. There is usually a history of the pain being affected by movement or by

115

inspiration, though these features may be present with intrathoracic disease as well.

The patient may sometimes give a history of trauma or strain, or of relation to emotional stress.

Examination may reveal tenderness, often over the costal cartilages. This is not an invariable finding since deeper tissues may be involved. If tenderness is over the sternoclavicular joint (Figure 10-1).

History, examination, and whatever laboratory tests are thought appropriate must be done to rule out cardiac, pulmonary, or gastroesophageal conditions.

In older patients, X rays or even a bone scan should be done to rule out bony lesions in the ribs.

Pathophysiology

Trauma, strain or overuse of the muscles from direct injury, carrying heavy objects, or coughing and sneezing may be responsible for muscular pain or for pain in the muscular insertions in the costal cartilages.

Stress can lead to chest muscle tightness just as it can to neckache or headache; often worry about chest pain creates a vicious cycle.

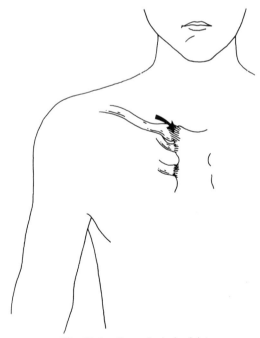

Fig. 10-1.   Sternoclavicular Joint.

Treatment

Reassurance is sometimes all that is needed.
The application of *wet heat* may be helpful.
Oral *antiinflammatory medications* can be prescribed.
In cases where psychological factors seem contributory, counselling and/or training in relaxation techniques may be of benefit.

Sternoclavicular Joint Pain (*Uncommon*)

Diagnosis

Tenderness is found over the joint (Figure 10-1).
There will be pain if the shoulder on the affected side is elevated more than one hundred degrees.

Pathophysiology

The joint may be strained acutely or can be painful secondary to degenerative change.

Treatment

Advise *heat, antiinflammatory medication,* and the *avoidance* of extreme shoulder elevation.

CHEST TRAUMA

Note: Open chest trauma or severe blunt trauma is beyond the scope of this discussion; the reader is referred to a text on emergency medicine.

Rib Bruise or Fracture (*Common*)

Diagnosis

There is a history of direct trauma and localized pain.
On examination, tenderness of the involved area is found; full breath sounds and a benign abdominal examination help to rule out pneumothorax or a ruptured viscus.
With a fracture, crepitus, ecchymosis, or palpable deformity may be found.

If there is any question of pneumothorax, a chest X ray must be done.
If the above complications are ruled out, rib X rays need to be done only:

If the patient requests one.

To help get an idea of how long disability will last (significantly longer if there is a fracture).

In older patients or in those in whom the force of injury seemed so slight in comparison to symptoms as to raise the question of a pathologic fracture.

Note that negative X ray does not rule out a costochondral separation.

Treatment

Treatment is the same whether a fracture or separation is present or not.

Sufficient *analgesia* should be supplied so that excessive splinting does not occur, because that can lead to atelectasis and pneumonia. For the same reason the patient should be told to take ten deep breaths an hour and to avoid smoking. This is especially important in older patients or those with underlying lung disease. For obvious reasons a rib belt should not be used.

11

The Low Back

Back pain in *children* and *adolescents*, as well as *abnormal curvature* of the back (including scoliosis), are discussed in Chapter 19. *Neurologic symptoms* and *diffuse pain* in the *lower extremities* are the subjects of Chapter 13. Further details on therapeutic suggestions made in this chapter can be found in Chapter 3. For the anatomy of the lumbosacral spine see Figure 11-1.

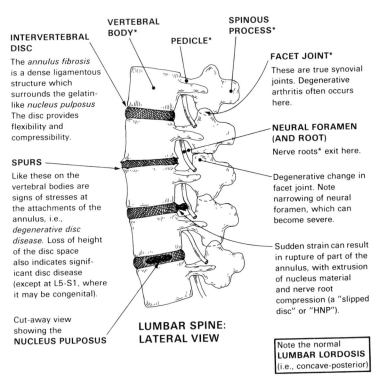

MUSCLES* are not shown.

INTERVERTEBRAL DISC
The *annulus fibrosis* is a dense ligamentous structure which surrounds the gelatin-like *nucleus pulposus*. The disc provides flexibility and compressibility.

VERTEBRAL BODY*

PEDICLE*

SPINOUS PROCESS*

FACET JOINT*
These are true synovial joints. Degenerative arthritis often occurs here.

NEURAL FORAMEN (AND ROOT)
Nerve roots* exit here.

SPURS
Like these on the vertebral bodies are signs of stresses at the attachments of the annulus, i.e., *degenerative disc disease.* Loss of height of the disc space also indicates signif-icant disc disease (except at L5-S1, where it may be congenital).

Degenerative change in facet joint. Note narrowing of neural foramen, which can become severe.

Sudden strain can result in rupture of part of the annulus, with extrusion of nucleus material and nerve root compression (a "slipped disc" or "HNP").

Cut-away view showing the **NUCLEUS PULPOSUS**

LUMBAR SPINE: LATERAL VIEW

Note the normal **LUMBAR LORDOSIS** (i.e., concave-posterior)

Fig. 11-1.   Anatomy of the lumbosacral spine. Asterisks indicate pain-sensitive structures.

EVALUATION

Always keep in mind that low back pain can be *referred* from other structures such as the kidneys, prostate, female pelvic organs, gastrointestinal tract, pancreas and abdominal aorta. The differential diagnosis of these conditions is beyond the scope of this book, but their possibility must not be forgotten, and appropriate history, physical examination and laboratory tests must be done to rule them out, especially if the clinical picture is not typical for musculoskeletal pain.

History

Ask about the *duration* and *location* of the pain.
Has there been any *trauma, strain,* or *overuse*?

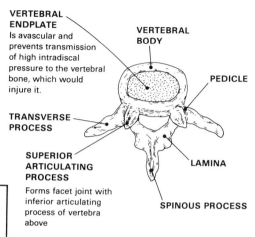

VERTEBRAL ENDPLATE
Is avascular and prevents transmission of high intradiscal pressure to the vertebral bone, which would injure it.

VERTEBRAL BODY

PEDICLE

TRANSVERSE PROCESS

SUPERIOR ARTICULATING PROCESS
Forms facet joint with inferior articulating process of vertebra above

LAMINA

SPINOUS PROCESS

Degenerative changes of the annulus fibrosis (secondary to mechanical stresses), along with loss of fluid content of the nucleus, lead to small malalignments between vertebrae; postural abnormalities lead to malalignments as well. These malalignments result in stress on the facet joints, with the subsequent development of degenerative arthritis.

A LUMBAR VERTEBRA* TOP VIEW

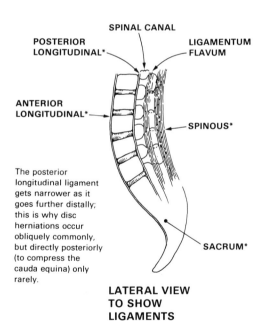

SPINAL CANAL

POSTERIOR LONGITUDINAL*

LIGAMENTUM FLAVUM

ANTERIOR LONGITUDINAL*

SPINOUS*

The posterior longitudinal ligament gets narrower as it goes further distally; this is why disc herniations occur obliquely commonly, but directly posteriorly (to compress the cauda equina) only rarely.

SACRUM*

LATERAL VIEW TO SHOW LIGAMENTS

Fig. 11-1. Continued.

Is there any *radiation* of pain into the legs?

Inquire about *weakness, numbness,* or *paresthesias* in the legs.

Find out about *previous back problems,* diagnoses, X rays, and treatments, including chiropractic.

What are *precipitating* and *relieving factors*?

Does the patient have any *urinary, gastrointestinal,* or *gynecologic* symptoms or *fevers, chills,* or *sweats*?

How does the pain *affect* the patient's life at home, at work, and at play?

Examination

Look for gross *postural changes* (Figure 11-2), which are commonly contributing factors in CHRONIC LOW BACK PAIN:

Hyperlordosis

Loss of the normal lumbar lordosis (which may be a chronic postural problem
    but more commonly is due to muscle spasm in acute decompensations)

A fixed kyphosis (implying ANKYLOSING SPONDYLITIS, the residue of
    SCHEUERMANN'S DISEASE, or the result of multiple compression
    fractures)

A lateral curvature (a fixed structural scoliosis will not disappear with recum-
    bency, while that due to muscle spasm usually will; the former will usually
    also be associated with deformity of the thoracic cage; see Chapter 19)

Excessive prominence of a spinous process which may indicate SPON-
    DYLOLISTHESIS

Examine *range of motion*:

Flexion can be measured by seeing how close the patient can get his fingers
    to the floor. Note if the patient flexes only one knee; this is a good indicator
    of sciatic nerve irritation. Putting fingers on adjacent spinous processes and
    noting how they separate during flexion may be helpful. (Restriction may
    be a clue to ANKYLOSING SPONDYLITIS).

See how far the patient can extend backward before the pelvis tilts.

Left and right lean can be measured by seeing how far down his thigh the
    patient can touch.

Look for *tenderness*.

Occasionally *trigger points* may be found, which are extremely tender nodules
    of muscle.

With the patient prone, anterior pressure or lateral movement of a spinous
    process that is especially painful may help to localize the segment responsi-
    ble for the patient's problem.

In young people with acute muscular strains, only the involved area will be
    tender. Tenderness over the sacroiliac joints can sometimes be related to

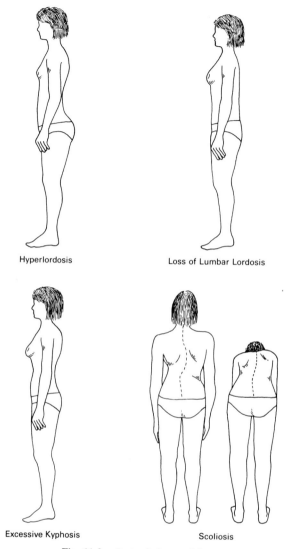

Hyperlordosis   Loss of Lumbar Lordosis

Excessive Kyphosis   Scoliosis

Fig. 11-2. Postural abnormalities.

pathology there. Other than these points, localization of tenderness is not helpful. (Tenderness in the sciatic notch area is usually referred.)

In chronic cases and in young patients, measure *leg length* (from each anterior superior iliac spine to its medial malleolus: see Figure 11-3).

Look for evidence of *nerve root irritation* by performing the *straight leg*

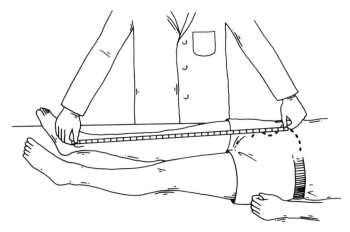

Fig. 11-3. Measuring leg length.

raising test (Figure 11-4). Note that *contralateral* leg pain with this test is *strong* evidence of root irritation. The *bowstring sign* is also quite specific for root tension. The *reverse straight leg raising* is positive with L4 root compression.

Problems of the *sacroiliac joint* are rare, but can be tested for by eliciting pain with lateral compression of the iliac crest or resisted abduction of the hip joint (Figure 11-5).

Finding a normal *range of motion* without pain or tenderness in the *hip* will

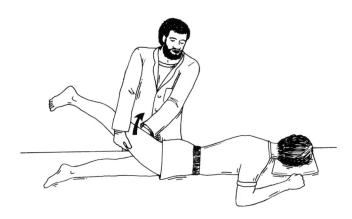

REVERSE STRAIGHT LEG RAISING:
This gives radicular pain in lesions of the L4 root.

Fig. 11-4. Radicular signs.

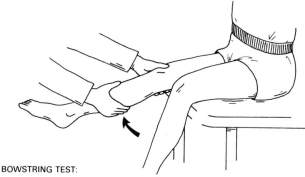

BOWSTRING TEST:
Fingers compress the tibial nerve in the popliteal space as the leg is straightened.

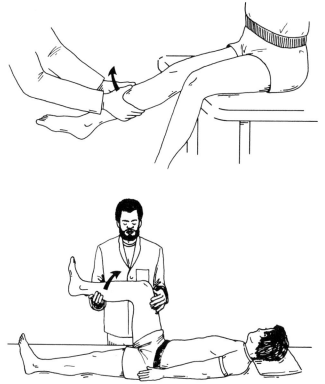

STRAIGHT LEG RAISING:
This is positive if it produces pain in the leg, or if *LeSegue's Modification* is positive. If the patient complains of back pain as the leg is being straightened, flex it a little bit to remove the pain and then see if the pain recurs with forced dorsiflexion of the ankle.

Fig. 11-4. Continued.

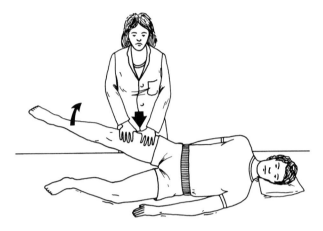

RESISTED HIP ABDUCTION:
Side with the suspected SI joint pathology is up.

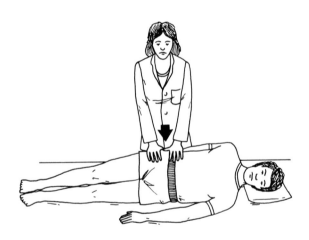

SI JOINT COMPRESSION

Fig. 11-5. Sacroiliac joint signs.

rule out that joint as the origin of pain; or *Patrick's test* (Figure 11-6) may be performed for the same purpose.

Test *strength* and *sensation* in the legs. See Figure 11-7 for the key muscle groups and sensory areas to test for lesions of the L4, L5, and S1 nerve roots. Check the patellar and achilles *reflexes*.

Feel the *abdomen* for tenderness or masses and listen for bruits; look for punch tenderness in the costovertebral angle area; and when the history is at all sugges-

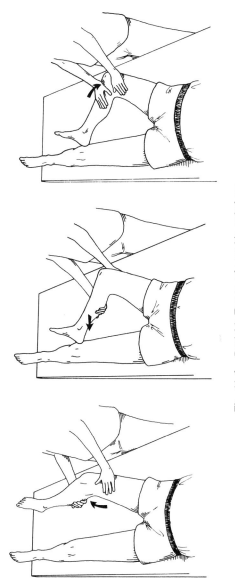

Fig. 11-6. Patrick's Test to rule out hip pathology.

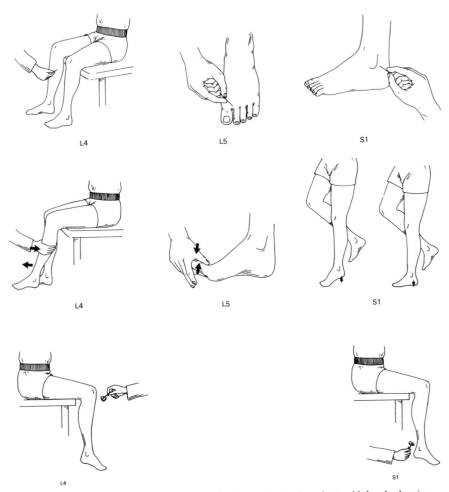

Fig. 11-7.   Screening lower extremity neurologic examination in patients with low back pain.

tive of a possible nonmusculoskeletal cause for the patient's pain, do a *rectal* examination and a *prostate* or *pelvic* examination. All this is of course to rule out back pain referred from vascular, urinary, gastrointestinal, or gynecologic struc-tures.

In CHRONIC LOW BACK PAIN, *psychologic* factors are often a prominent component, or at least have a major role in determining how the patient reacts to his or her pain, and so an evaluation of the patient's mental and emotional status is in order.

X Ray

Although when X rays should be taken was discussed in Chapter 2, the situation is somewhat more complicated in the case of the low back for two contradictory reasons. On the one hand, because of the amount of soft tissue that has to be transversed, low back X rays involve *more radiation exposure,* and the exposure is to a variety of particularly radiosensitive tissues (the gonads and the kidneys included). On the other hand, certain *very serious conditions* such as osteomyelitis or tumor will be easily missed for a significant period of time if the criteria for taking X rays are too stringent, and other lesions that may affect the treatment course of mechanical low back pain will not be diagnosed. Therefore, the following criteria for X raying the lumbosacral spine are suggested:

A history of direct trauma.

A history of indirect trauma with immediate severe pain.

Any objective neurologic findings.

Fever, chills or sweats without other obvious source.

History or examination suggestive of sacroileitis (for example, a family history of ANKYLOSING SPONDYLITIS, a personal history of psoriasis or colitis, or positive sacroiliac joint tests. In such cases order sacroiliac joint views).

Examination suggestive of spondylolisthesis.

Any patient under 25 without obvious history of strain or with pain persisting for more than two weeks.

Any patient over 50 or with a past history of malignancy.

Any case in which pain is prolonged (i.e., does not resolve as expected).

A patient who needs the reassurance of a normal X ray.

A *bone scan* should be considered the next step in cases in which the patient's age, past history, and current symptoms or examination are suggestive of the possibility of infection, inflammation or neoplasm and in whom X rays are negative.

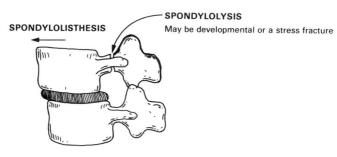

Fig. 11-8. Spondylolisthesis and spondylolysis.

Note that the following X ray findings, though thought to be etiologic factors of low back pain in the past, are now considered to be no more than incidental findings in most cases, though some predisposition to degenerative disc disease is possible.
Spina bifida
Hemivertebra
Iliotransverse joint
Lumbarization of S1
Sacralization of L5

Other Lab Tests

See the discussion in Chapter 2. Note, though, that the *ESR* is an extremely useful test in differentiating mechanical lower back pain from that arising from primarily inflammatory lesions such as infection or spondylitis.

Of course, a *urinalysis* will often be necessary to rule out a urinary tract cause for the patient's pain, especially in females.

The HLA-B27 marker has a much higher incidence in patients with spondylitis.

Other lab tests used, especially in older patients, to rule out tumor are CBC, calcium, phosphate, alkaline phosphatase, acid phosphatase, and serum protein electrophoresis.

LOW BACK PAIN AND STRAIN

Acute Low Back Strain (*Extremely Common*)

Diagnosis

Localized or diffuse low back pain occurs following strain or overuse.
The pain may radiate to the buttock or thigh.
There is no radiation beyond the knee and no neurologic symptoms or findings.
Tenderness and pain on range of motion are variable.
See Table 11-I.

Pathophysiology

This syndrome can be due to any of the following:
Muscle strain and/or spasm.

Table 11-I   Examination of Patients with Acute Back Pain

1. Look for tenderness.
2. Record range-of-motion: flexion, extension, left and right lean.
3. Do a straight-leg-raising test (Figure 11-4).
4. Do a screening neurologic exam (Figure 11-7).
5. Consider and if necessary exclude by appropriate means a nonmusculoskeletal cause for the patient's pain.

Strain on facet joints from misalignment of vertebrae, usually secondary to disc degeneration.

Ligamentous strain.

A bulge or herniation of an intervertebral disc not large enough to cause significant nerve root impingement.

Treatment

The patient should *rest* the back, with progressive ambulation as symptoms abate (see Patient Handout 11-1).

Local *heat* can be helpful.

Analgesic medication is important not only for the patient's comfort but also to help break the pain → muscle spasm cycle.

If trigger points are found, *injection* may be helpful. See Figure 4-2.

If there is evidence of deconditioning (weak abdominal muscles, tight hamstrings, poor posture), *rehabilitative exercises* should be begun when the episode is completely resolved. *Postural instruction* should be given to help avoid recurrence. See Patient Handouts 11-1 through 11-4.

Acute Lumbosacral Radiculitis (Slipped Disc, Herniated Nucleus Pulposus) (*Common*)

Diagnosis

Usually there is sudden onset of low back pain, often lateralized, often severe, with radiation down past the knee.

The inciting motion may not seem at all excessive.

Often the patient complains of paresthesias in the distribution of the involved root, and sometimes complains of lower extremity weakness as well.

Note that sometimes back pain is absent or minimal, and lower extremity neurologic symptoms are the main manifestation.

On examination, tenderness and restriction of range of motion is variable.

The straight leg raising and/or other nerve root tests will be positive (Figure 11-4).

Often there is decrease or loss of the involved deep tendon reflex.

Often there is decreased sensation in the dermatome of the involved nerve root (see Figure 13-2).

Sometimes there is weakness in muscles supplied by the involved root (see Figure 13-2).

Pathophysiology

Degenerative changes in the annulus fibrosis lead to its bulging or to its actual rupture with extrusion of nucleus pulposus material, irritating and often compressing nerve roots.

Treatment

Strict bedrest should be prescribed with *progressive ambulation* as symptoms resolve (Patient Handout 11-1).

Antiinflammatory medication is advisable to decrease the inflammation that invariably occurs around the involved root. In severe cases, high-dose corticosteroids with a rapid taper have been used to good effect.

Sufficient *analgesia* is important to break the pain cycle.

Occasionally a short course of an *antianxiety agent* may be necessary, especially if a previously active person must be restricted to bed.

Any *bowel* or *bladder dysfunction* or *rapidly progressive leg weakness* implies a cauda equina lesion and a *surgical emergency,* and the patient must be seen by a neurosurgeon or orthopedist *immediately.*

Progressive sensory loss or any *significant weakness* is an indication for urgent consultation. (Further tests such as myelography may be necessary to confirm the diagnosis and rule out a spinal cord tumor or hematoma.)

Failure to improve on a strict regimen of bedrest after a week or two is another indication for referral for further evaluation or other treatment modalities (such as traction).

Once the acute episode is resolved, *postural instruction* should be given to help avoid recurrence (Patient Handout 11-2). If there is evidence of deconditioning (weak abdominal and back muscles, tight hamstrings, poor posture), *rehabilitative exercises* should be prescribed as well (Patient Handout 11-3).

Recurrent Low Back Pain *(Extremely Common)*

Diagnosis

The patient gives a history of recurrent acute episodes with or without neurologic symptoms or significant radiation.

Occasionally back pain is minimal or absent and the lower extremity neurologic symptoms are predominant.

Pathophysiology

The most likely theory is that disc degeneration leads to chronic malpositioning and strain on the facet joints, allowing relatively minor episodes of strain and overuse to "push the facet joint too far" into a painful position.

Of course, pain may also be due to recurrent muscular or ligamentous strains due to deconditioning.

Some episodes may represent actual disc bulge or rupture.

As the years go on, these episodes often merge into CHRONIC LOW BACK PAIN.

Treatment

Treat the acute episode as in ACUTE BACK STRAIN.

Postural advice and *rehabilitative exercises* are all-important. *Weight loss* is critical in the obese.

Chronic Low Back Pain (*Extremely Common*)

Diagnosis

An outline for examination of patients with recurrent or chronic low back pain is presented in Table 11-II. Look specifically for any of the following contributing factors:

Leg length discrepancy

Hyperlordosis

Deconditioning and poor posture

Loss of lumbar lordosis

Trigger points

Tension, depression, secondary gain

Significant degenerative changes in the facet joints seen on X ray

Evidence of nerve root compression

Pathophysiology

The most common mechanism is probably disc degeneration leading to malalignment, strain and eventually degeneration of the facet joints.

Deconditioning and poor posture may make that worse, or may cause pain by stresses on other pain-sensitive structures.

Table 11-II   Outline for Examination of Patients with Recurrent or Chronic Low Back Pain

After obtaining the history, with the patient undressed and seated,

1. Observe the patient's posture.
2. Do a straight-leg-raising test (Figure 11-4).
3. Check the patellar and achilles reflexes.
4. Check pin sensation at the medial lower leg (L4), the first web space (L5) and just in front of the lateral malleolus (S1).
5. Check strength of knee extension (L4) and dorsiflexion of the distal great toe (L5) against your resistance.
6. Check vascular status in the feet.

Then have the patient stand up.

7. Again, observe the curvature of the spine.
8. Have the patient go up and down on his toes six times on each foot (S1 strength).
9. Observe the range of flexion, extension and left and right lean.

Have the patient lie down supine.

10. Examine the abdomen for tenderness, masses, and bruits.
11. Measure leg lengths.
12. Perform Patrick's maneuver (Figure 11-6) and repeat the straight-leg-raising test.

Have the patient lie on his or her side (painful side up).

13. Test for SI joint problems with resisted hip abduction (Figure 11-5).

Have the patient roll to a prone position.

14. Do the reverse straight-leg-raise.
15. Look for tender spots, and look for pain with the application of pressure to, and the side-to-side rocking of, each spinous process.

Perform rectal/prostate or pelvic/rectal examination.

With some practice the entire examination (excluding the pelvic) will take no more than a few minutes. Of course, additional examination should be done as indicated.

Note that nerve root compression may not always be by the disc. Other mechanisms that can be responsible include:

Impingement on the foramen by a subluxing facet

Compression of the nerve root between a hypertrophied facet joint and the vertebral body

Kinking of the root around a pedicle

Psychogenic and environmental factors often become of paramount importance. For example:

Loss of income

Secondary gain (financial or through family or employer manipulation)

Fear of disability

Psychogenic muscle tension often is a prime component of the problem, especially in younger patients

Depression may be a primary or secondary factor.

Treatment

Of course, significant or progressive radicular symptoms should be referred to the specialist for further workup (myelography to detect a possible cord tumor, for example) and aggressive therapy.

Psychogenic factors must be addressed:

By the use of *relaxation techniques* (the one outlined in Patient Handout 4-3 or others)

By referral for *counseling* and/or psychotherapy where appropriate

By the use of *antidepressant medications* where appropriate

Awareness must be kept of the patient's fears, family and financial problems, and secondary gain factors

Though psychologic factors are of great importance, the pain is usually quite real and should be dealt with:

Antiinflammatory medications are helpful if there is posterior joint degeneration or root compromise.

Postural training and *rehabilitative exercises* are very important (Patient Handouts 11-2 and 11-3).

If there is a loss of lumbar lordosis, an *extension exercise program* (Patient Handout 11-4) may be more helpful than the standard exercises in Patient Handout 11-3. Note that it is *contraindicated* with any evidence of root compression.

Weight loss is to be emphasized to the obese patient.

General *aerobic conditioning* is often very helpful from both physical and psychologic points of view. A program should be prescribed taking into account the patient's condition; it can involve walking, swimming, cycling (possibly stationary cycling at home), or running.

Mild analgesics such as aspirin may be advised not only for patient comfort but also to help break the pain → spasm → strain → pain cycle.

Physical measures such as wet heat or ice massage may help.

If there is a leg length discrepancy of greater than one cm, correction with a lift of about one half the discrepancy is advisable.

Referral to a physical therapist, physiatrist or orthopedist should be carried out in cases that are not responding well for consideration of other modes of therapy (facet joint injection, facet rhizotomy, bracing, manipulation, etc.). Surgery should be considered a last resort, except of course in cases where there is a significant or progressive neurologic deficit. But last resorts sometimes need to be resorted to.

Spondylolisthesis (*Uncommon*)

Diagnosis

Forward slippage of a vertebra on the one below it (most commonly L5 on S1) is noted on X ray (Figure 11-8).
It can cause low back pain and/or nerve root compromise.

Pathophysiology

Most commonly spondylolisthesis is due to spondylolysis (which itself can be a developmental defect or a stress fracture), but can rarely be due to isthmic stretching in a child, to trauma, or itself be a congenital defect.

Treatment

Many cases are asymptomatic. The chance of a spondylolisthesis seen on X ray being the etiology of the patient's symptoms is probably greater the younger the patient is.
In spondylolytic spondylolisthesis, methods of treatment are fairly similar to those in other etiologies of low back pain. Flexion exercises (Patient Handout 11-3) are probably of greater importance, however, and patients with fairly severe symptoms may be helped by a properly fitted brace.
The management of the other forms of spondylolisthesis and of severe cases of spondylolytic spondylolisthesis should be left to the specialist.

Ankylosing Spondylitis and Sacroileitis (*Uncommon*)

Diagnosis

Low back pain occurs in a young male.
Examination shows positive sacroiliac joint tests, limitation of forward flexion and a limitation of chest expansion (because of involvement of the costovertebral joints).
X ray shows sacroiliac joint changes (blurring of the margins, irregular erosions and sclerosis*) and syndesmophytes (distinguished from the osteophyte of degenerative disc disease by the fact that the former extend more vertically).

---

*Not to be confused with OSTEITIS CONDENSANS ILII, a radiologic finding of no significance, usually in multiparous women, in which sclerosis is confined to the illiac side of the joint. This condition is analogous to the equally insignificant OSTEITIS PUBIS (sclerosis of the pubic symphisis), found in runners, soccer players, and horseback riders as well as mothers.

An elevated ESR and a positive HLA-B27 antigen are found in 90% of cases.
About one-fourth of patients have large peripheral joint involvement, usually of the hips and shoulders.
About one-fourth of patients have iritis at some time in the course of their disease.
A small percentage of patients develop aortic insufficiency.
Similar syndromes may be seen in patients with psoriatic arthritis or the arthritis associated with inflammatory bowel disease. Back symptoms in patients with Reiter's syndrome are usually not significant. See Chapter 23 for further discussion of these conditions.

Treatment

Use *antiinflammatory* medication.
Postural instruction to minimize the eventual deformities by maintaining a straight spine is critical.
Referral to a physical therapist for long-term follow-up is wise.
Patients with cardiac or eye involvement should be followed by specialists in those fields.

Sacroiliac Joint Sprains *(Rare)*

Diagnosis

These are *extremely rare* except following childbirth or with trauma to females in the late stages of pregnancy.
Examination reveals tenderness over the sacroiliac joints and positive sacroiliac joint tests.

Treatment

Use bedrest, antiinflammatories, and analgesics.

Osteoporosis *(Very Common,* though it is unclear how often it is a significant contributing factor in low back pain)

Diagnosis

The patient usually reports dull, nagging, boring pain that increases with activity and decreases with rest.
The course is often punctuated by episodes of severe pain corresponding to

compression fractures, or the latter episodes may comprise the only symptoms, and the patient may be pain-free in the interim.

X ray reveals loss of horizontal trabeculae (with a secondary relative prominence of vertical striations), decreased bone density (hard to quantitate accurately except by computerized tomography), and possibly evidence of old compression fractures of the vertebral bodies.

Serum calcium, phosphorus, alkaline phosphatase, CBC, thyroid function tests, electrolytes, and SPEP must be obtained to rule out other causes of decreased bone mass such as osteomalacia or otherwise inapparent Paget's Disease, as well as to rule out secondary osteoporosis from such entities as hyperparathyroidism, Cushing's syndrome, metastatic cancer, and myeloma.

Pathophysiology

Loss of bone mass is due to excess bone resorption over creation, and is seen in older patients and postmenopausal women. The cause is unknown in most cases.

It can sometimes be secondary to other conditions which must be ruled out with the tests mentioned above.

The pain of osteoporosis is most likely due to increased venous pressure within the vertebra, secondary to increased blood volume (since bone mass is decreased and it is known that fat volume is unchanged).

The theory that microfractures are responsible is refuted by the fact that the vertebral cortex is probably not pain-sensitive. (Clinical compression fractures produce pain by stresses on the posterior joints and other pain-sensitive structures.)

Treatment

Maintenance of activity is of the utmost importance; inactivity leads to bone loss even in young people.

Estrogens given at menopause can retard bone loss, though certainly not reverse it. Whether this benefit of the hormone is worth the possible increased risk of thromboembolic phenomena and breast cancer (and the definite risk of endometrial carcinoma in women who still have their uterus) is an open question.

Dietary *calcium* supplementation in doses of about a gram of elemental calcium per day probably helps retard bone loss, though not as well as estrogens. *Vitamin D* supplementation may help as well.

Treatment with *fluoride* unquestionably increases radiologic bone density, but whether there is any real increase in bone strength, or just an artifact caused by the fact that fluoride is very radioopaque, is still controversial.

Osteomyelitis and Discitis (*Rare*)

Diagnosis

To make the diagnosis, the condition must be suspected, since often fever and other constitutional symptoms are absent, and X ray changes may not become apparent for weeks.

The ESR is invariably elevated, and should be performed in any patient whose course is at all prolonged or otherwise atypical, as well as in high-risk groups (patients with recent pelvic infection or back surgery, IV drug users, children and adolescents with back pain).

A bone scan will be positive before evidence of infection can be seen on X ray.

Pathophysiology

Vertebral osteomyelitis is usually secondary to staph aureus or, in needle users, to gram-negative organisms; the bacteria most commonly reach the spine by hematogenous spread.

Disc space infection (except in postlaminectomy patients) is most common in children and teenagers and is thought to be viral.

Treatment

These patients should be *referred* to an orthopedist and an infectious disease specialist for management.

Benign Neoplasms (*Rare*)

These are usually found in patients under 30.

The pain is not necessarily relieved by rest.

The diagnosis is by X ray and/or bone scan.

These tumors are almost always found in the posterior elements of the vertebra.

Osteoid osteoma is the most common type; others are osteoblastoma, eosinophilic granuloma, aneurysmal bone cyst, and giant cell tumor.

The patient should be referred to an orthopedic surgeon.

Primary Malignant Neoplasms (*Uncommon*)

These usually occur in patients over 40.

Again, pain is not necessarily relieved by rest.

These tumors involve the vertebral bodies.

Myeloma is most common: X ray may show lytic lesions or just diffuse osteoporosis. Diagnosis is by abnormal serum protein electrophoresis; the ESR is invariably increased but is not specific. Patients with myeloma may have constitutional symptoms as well. See an internal medicine or hematology text for further details.

Chordoma is the only other primary malignant tumor of the spine seen at all commonly.

Patients should of course be referred to appropriate specialists.

Metastatic Tumors (*Common*)

The spine is the most common site of metastatic spread in the skeleton.

Only a minority of vertebrae shown to be involved at autopsy demonstrate lesions on X ray.

In a fair number of patients with malignancy, back pain secondary to metastasis will be the presenting symptom.

Therefore, *patients over 50* presenting with back pain, *especially* if without known trauma or strain and/or if unrelieved by rest, should be *considered to have a spinal metastasis until proven otherwise.*

If there is any level of suspicion and X ray is normal, a bone scan should be done as well as a CBC, ESR, SPEP, calcium, phosphorus, alkaline phosphatase, and acid phosphatase.

Most lesions are osteolytic, especially colon and thyroid carcinoma and hypernephroma. Breast, lung, and prostate cancer metastases can be osteoblastic.

The patient should of course be referred to the appropriate specialist for management.

Fractures and Dislocations

Their management is beyond the scope of this book and is best left to the orthopedist. Note should be taken of the fact that fractures of the transverse processes, though often appearing insignificant and treated as such, are associated with a tremendous amount of muscular and ligamentous damage and can result in significant and prolonged pain and disability.

Patient Handout 11-1

What To Do When You Strain Your Back

The most important thing you can do for any injured part of your body is to rest it, and the back is no exception. The way you rest your back is to remain lying down in a comfortable position on a firm surface (use the carpet or put a bed-board under your mattress if your bed is soft). You will probably find that one of the positions shown is most comfortable.

How many hours a day you need to be lying down, and how many days you need to continue with it, depends simply on how much your back hurts and how long it takes to get better. But a word of warning: most people underestimate tremendously how much rest the back needs. The most common cause of recurring back pain requiring return visits to the doctor is the lack of enough rest. An hour or two a day just does not do it. If you have severe back pain, you need to be lying down 24 hours a day (except only for going to the bathroom).

As your back improves, you may begin getting less rest and walking around a bit, returning to bed when it begins to ache. As the days go on, you will be spending more time walking and less time lying down. But caution: sitting and standing still are bad for your back. Do not do them until your back is completely well again.

In addition to rest, you should relieve the pain by taking a couple of aspirin or acetamenophen tablets every 4 to 6 hours. These not only make you feel more comfortable but also help to stop the extra muscle spasm that pain itself can cause. If your pain is not controlled by these, let your doctor know.

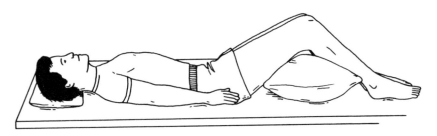

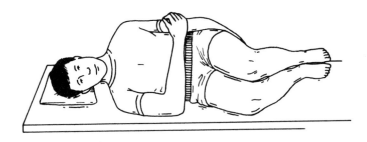

Putting a heating pad or hot towel on the most sore area for twenty minutes every few hours can help too.

Your doctor may have given you some tips to help prevent this from happening again. Follow them. He or she may also have prescribed some exercises. They are important. Wait until you have felt all better for a couple of weeks, and then start them as directed.

If you develop numbness or weakness in your legs at any time, be sure to let your doctor know. And if you develop any problems with your bowel or bladder control (this is very, very rare), call your doctor right away or go to an emergency room if you cannot reach him or her.

If you follow the tips in this handout, you will be back in action as soon as possible. If you do not, your pain can go on much longer than necessary.

Patient Handout 11-2

Taking Care of Your Back

Here are some tips on preventing strain to your back. In addition to these measures, you should do whatever else your doctor has recommended for your specific back problem.

1. A GENERAL RULE
   Never stay in one position for a long time; move around frequently.
2. SLEEPING
   Be sure your mattress is firm and doesn't sag in the middle. Put a bed board between the mattress and box spring if necessary.

WRONG

WRONG

RIGHT

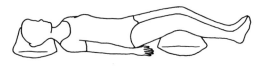

RIGHT

Either lie on your side with your legs curled up, or sleep on your back with a pillow under your knees. Never sleep on your stomach, nor on your back with your legs straight out.

3. SITTING
    Don't sit on a chair that is too high.

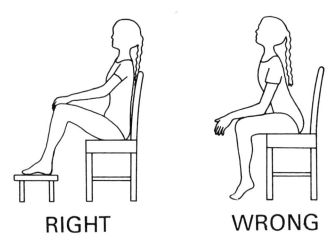

RIGHT   WRONG

Put your feet on a stool so that your knees are bent. If nothing is available to put your feet on, cross your legs (change legs every so often).

4. STANDING
    Stand tall, with your back flat and your knees relaxed. Don't stand with your knees stiff and your low back arched.

RIGHT   WRONG

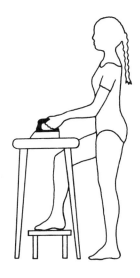

RIGHT   WRONG

If you have to stand for a prolonged period, alternately put one foot and then the other up on a footrest (when ironing, for example).

5.   DRIVING
Move your seat up so that you're sitting closer to the steering wheel and pedals. Remember to wear your seat and shoulder belts.

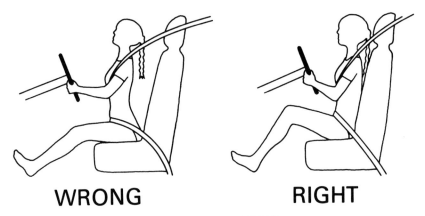

WRONG   RIGHT

6.   LEANING, BENDING, AND REACHING
Use a stepladder to avoid unnecessary reaching up.

RIGHT   WRONG   RIGHT

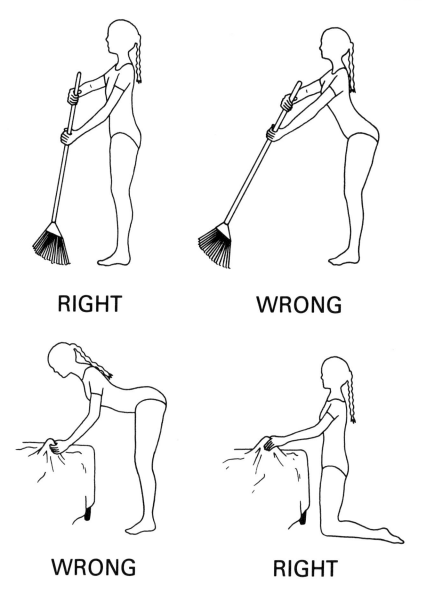

RIGHT   WRONG

WRONG   RIGHT

Never, ever bend over without bending your knees and tucking your buttocks under.

Don't reach forward to vacuum, rake etc. Hold the handle close to your body.

Kneel to make the bed instead of bending over.

7. LIFTING AND CARRYING
   Don't twist to lift something or put it down: Turn your feet instead.
   Bend at the hips and knees, never at the waist.

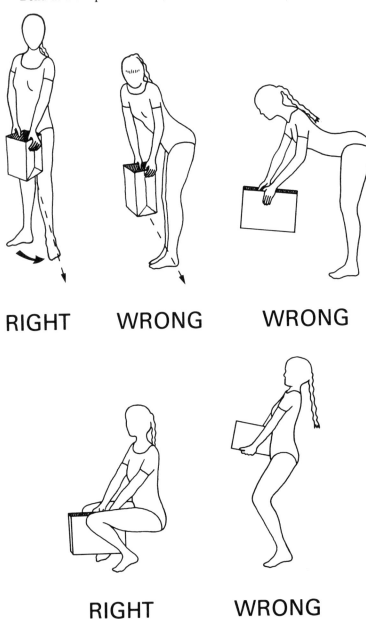

RIGHT   WRONG   WRONG

RIGHT   WRONG

WRONG   RIGHT   WRONG

WRONG   RIGHT

Don't handle things that are too heavy.
Hold objects close to your body, not with your arms extended out.
Don't carry anything unbalanced.
Never carry anything above waist level.

8. DRESSING

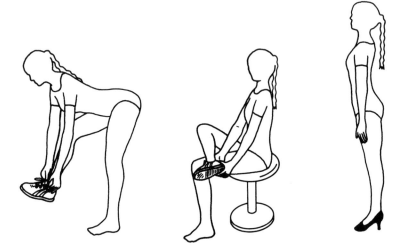

WRONG   RIGHT   WRONG

Don't bend over to put on your socks and shoes: sit down instead.
Don't wear high heels.

9. COUGHING AND SNEEZING

RIGHT

Round your back and bend your knees if you think you are going to
cough or sneeze.

10.   EATING
      If you are overweight, losing weight is the single most important thing
      you can do for your back.

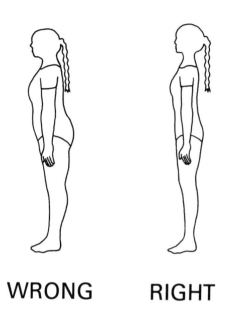

WRONG   RIGHT

Patient Handout 11-3

Back Exercises

These exercises are designed to prevent excessive swayback and to strengthen the abdominal muscles that help take a load off your back. Don't do them while your back is hurting; wait until it has felt all better for a while. If any of these exercises cause pain, stop doing them and let your doctor know.

In order to be effective, you have to do these exercises every day (twice a day is best). Doing them only once in a while is worse than not doing them at all.

Applying some heat to your low back before you start your exercises may help a bit. Do *three* repetitions of each exercise the first time. Increase the number of repetitions by one every few days until you are doing ten repetitions of each one. Do them slow and steadily, not jerkily.

1. RELAXING
   Lie on a firm, comfortable surface (like the carpet) with your back flat and your knees bent. Breathe in slowly for a count of two and then out for a count of two. Keep breathing in this slow rhythm as you tighten your fists and then let them relax, feeling the relaxation spread up your arms and into your neck, and then down to your back and legs.

2. PELVIC TILT
   Now tighten your stomach and flatten your low back to the floor. Hold for a slow count of five before relaxing. Be sure the soles of your feet stay flat on the floor while doing this exercise.

3. KNEE TO CHEST
   Now grab one knee with your hands and pull it slowly toward your chest. Pull firmly for a slow count of five. Put it down slowly and do the same thing with the other knee.

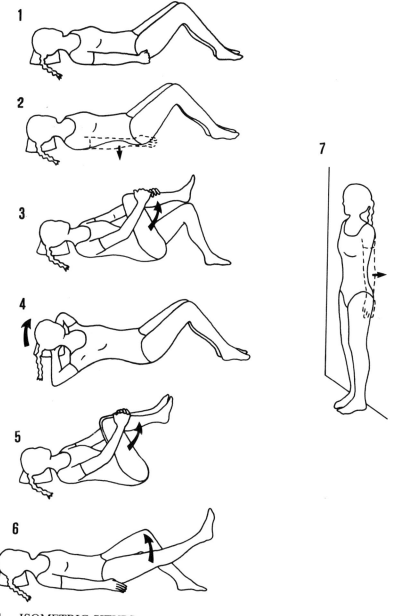

4. ISOMETRIC SITUPS

Keep your feet flat on the floor and your knees bent as you place your hands behind your head. Now tighten your stomach muscles as if you were trying to do a sit-up, but don't allow your head to lift more than a few inches. Hold for a slow count of five and relax.

5. DOUBLE KNEE TO CHEST

Same as knee to chest, but do both knees together.

6. LEG RAISE

With your knees bent and your feet flat on the floor, flatten your low back to the floor. Now straighten one leg and raise it as far as you can. Hold for a slow count of five before lowering it. Do the same with the other leg. DON'T DO THIS ONE IF YOU HAVE WEAKNESS OR NUMBNESS IN YOUR LEGS.

7. STANDING PELVIC TILT

Now get up and stand with your back against a wall. Flatten your low back against the wall.

Hold for a slow count of five and relax.

By doing these exercises faithfully you can save yourself a lot of suffering later.

Patient Handout 11-4

Back Care and Exercises (Extension)

Your doctor has determined that your back problem is at least in part due to loss of the normal curve (lordosis) in your low back. That is, your low back is too flat.

One thing you can do about this is to be aware of the need to maintain the hollow in your low back. This is especially important when you are sitting for prolonged periods, as in a car or while reading or watching TV. Either learn to

use your own muscles to keep a proper hollow in your low back or use a small pillow there. A rolled-up towel does quite nicely.

There is also a very simple exercise that you should do 5 to 10 times twice a day, and whenever else your low back begins to feel sore. Simply stand, put your hands on your hips, and arch your low back backwards for a slow count of five before relaxing. It is especially important to do this during activities in which you are bent over a lot. If you have (or develop) pain down your legs or weakness or numbness in your leg, don't do this exercise; instead, let your doctor know.

12

The Coccyx

EVALUATION

What is the *duration* of the symptoms?

Was there any *trauma,* or was the onset of pain related to childbirth?

Are there any *gastrointestinal* or *gynecologic* symptoms?

Tenderness and *mobility* of the coccyx should be evaluated by performing a digital rectal examination as well as palpating the external aspect.

The *lumbosacral* and *thoracic spine* should be palpated as well.

Be very careful, especially with acute pain, that what you are dealing with is not a *perirectal abscess.*

COCCYX BRUISE OR FRACTURE (*Common*)

Diagnosis

There is usually a history of a fall on the tailbone. The coccyx can sometimes be fractured during childbirth.

Examination reveals coccygeal tenderness, and if a fracture is present, painful excessive mobility of the coccyx on rectal examination.

If palpation of the more proximal spine is painful, an X ray of the sore area must be done to rule out an associated compression fracture there.

Treatment

A *ring seat* (a life preserver serves quite well) takes pressure off the injured part.

Analgesics should be used as needed.

The presence of a fracture means that pain will probably persist for months, but no additional specific treatment is indicated or available.

Reminder: Acute coccyx area pain is most commonly due to a perirectal abscess.

COCCYDYNIA (*Uncommon*)

Diagnosis

The patient gives a history of weeks or months of pain, worse after sitting.

Remember that acute coccyx area pain is most commonly due to a *perirectal abscess.*

Be sure that there are no symptoms suggestive of gastrointestinal or gynecologic disease.

Examination reveals tenderness to palpation only. (Rectal exam is otherwise normal.)

If no tenderness is found, rule out the possibility of pain referred from gynecologic or retroperitoneal structures or the lumbosacral spine.

Pathophysiology

This is usually due to previous trauma or to frequent and/or prolonged sitting on hard surfaces.

Treatment

The patient should *avoid sitting on hard surfaces.*

A *ring seat* is helpful.

Steroid injection may be offered. Infiltrate a cc each of steroid and of local anesthetic around the tender area after sterile prep and ethyl chloride spray. Be sure no infection is present. See p. 23 for further details and precautions.

13

Neurologic Symptoms and Diffuse Pain of the Lower Extremity

EVALUATION

Neurologic symptoms and especially pain of the lower extremity can be a manifestation of *vascular disease*. The diagnosis and management of this problem is beyond the scope of this book, but its possibility must be kept in mind and appropriate history taken and examination and tests performed to exclude it.

Also, weakness, numbness, and paresthesias obviously are often due to lesions of the *brain* or *spinal cord*. A *polyneuropathy* will usually manifest itself as sensory loss in a "stocking" distribution without associated pain, but metabolic disease (especially diabetes) can affect single large nerves and lead to confusion with the compression neuropathies discussed herein. And diseases of *muscle* or

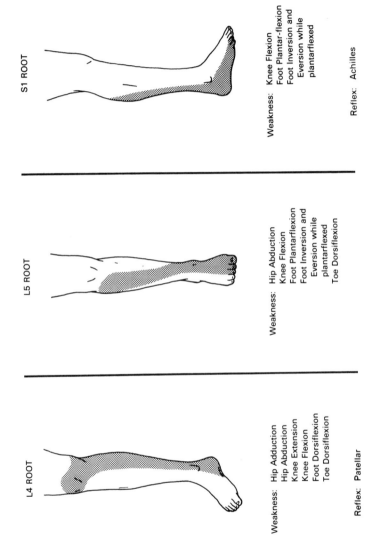

L4 ROOT

Weakness: Hip Adduction
Hip Abduction
Knee Extension
Knee Flexion
Foot Dorsiflexion
Toe Dorsiflexion

Reflex: Patellar

L5 ROOT

Weakness: Hip Abduction
Knee Flexion
Foot Plantarflexion
Foot Inversion and
Eversion while
plantarflexed
Toe Dorsiflexion

S1 ROOT

Weakness: Knee Flexion
Foot Plantar-flexion
Foot Inversion and
Eversion while
plantarflexed

Reflex: Achilles

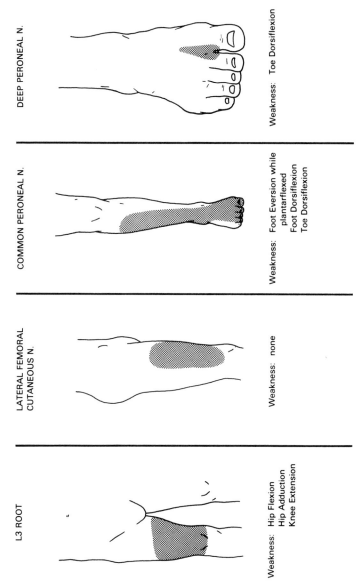

L3 ROOT

Weakness: Hip Flexion
Hip Adduction
Knee Extension

LATERAL FEMORAL
CUTANEOUS N.

Weakness: none

COMMON PERONEAL N.

Weakness: Foot Eversion while
plantarflexed
Foot Dorsiflexion
Toe Dorsiflexion

DEEP PERONEAL N.

Weakness: Toe Dorsiflexion

Fig. 13-1. Common root and peripheral nerve syndromes (lower extremity). Syndromes are usually incomplete. Dermatomes and innervations can vary in some patients.

the *neuromuscular junction* can underlie motor weakness. Again, discussion of these conditions is well beyond the scope of this volume; but it is essential that whatever history, examination and tests are necessary to rule out such an etiology be performed whenever the initial history or examination are at all suggestive of such a possibility.

Occasionally neurologic symptoms and diffuse pain of the lower extremity can be due to nerve root impingement by *pelvic* or *retroperitoneal pathology*, a point that should be kept in mind.

History

Ask about the *duration* of symptoms and the *location* of *pain* and *paresthesias*. Often the patient's response to this question is too vague to be useful. See Figure 13-1 (pp. 160–161).

Has the patient noticed any *weakness* or *gait difficulty?*

Inquire about *precipitating factors*. Aching pain which comes on with a fairly consistent amount of walking and is then relieved by rest is of course suggestive of *intermittent claudication* due to vascular compromise. If vascular examination is normal, the patient may have the so-called *pseudoclaudication syndrome*, which is usually due to narrowing of the spinal canal from degenerative changes or tumor. The patient certainly needs a detailed examination, probably needs a myelogram, and very possibly will require surgery rather soon. A less consistent

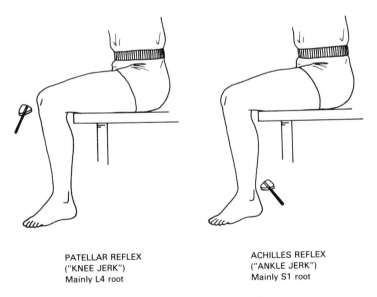

PATELLAR REFLEX
("KNEE JERK")
Mainly L4 root

ACHILLES REFLEX
("ANKLE JERK")
Mainly S1 root

Fig. 13-2.   Reflexes: lower extremity.

pattern of exacerbation with activity may be seen with pain referred from nerve root irritation in the back or an intermittent *compartment syndrome*.

Find out about any *back* or *buttock pain* and its association with the more distal symptoms.

Be sure to ask about any *weakness, numbness,* or *paresthesias* elsewhere in the body, or any headaches, visual or auditory symptoms, vertigo, imbalance, incoordination, speech difficulty, episodes of loss of consciousness, or recent head trauma. Any of these may of course suggest a central nervous system lesion.

Examination

Examine the *back* and do the *straight leg raising* test and other tests as necessary to see if the symptoms are reproduced or exacerbated. See Chapter 11 for a description of these maneuvers. Often distal pain and/or paresthesias are the only manifestations of nerve root compromise in the low back, back pain being conspicuous by its absence. As a matter of fact *this etiology is much more common than any of the other syndromes discussed in this chapter.* If such is the case, management is as discussed in Chapter 11 under the headings ACUTE LUMBOSACRAL RADICULITIS, RECURRENT LOW BACK PAIN, or CHRONIC LOW BACK PAIN.

Examine the *hip, knee,* and *ankle* for range of motion, swelling, and tenderness, since often a specific articular or periarticular condition will be the cause of the patient's complaints even though he or she is unable to localize the symptoms.

Look for *tenderness* and *tenseness* in the lower leg, and if present see if the pain is exacerbated by passive stretch of the involved compartment. If so the patient may have a COMPARTMENT SYNDROME which if acute is a true orthopedic emergency, requiring STAT referral.

Check distal *pulses* and *capillary filling*.

Elicit and knee and ankle *reflexes* (Figure 13-2).

Examine *strength*. See Table 13-I for various motions to be tested and their root and peripheral innervations. *Significant* or *progressive weakness* is always an indication for referral for more precise diagnosis (e.g., EMG) and more aggressive management, and may be suggestive of a CNS lesion.

Look for muscle *atrophy*.

Test for *pin sensation*. When a more sensitive examination needs to be done, check *two-point discrimination*. See Figure 13-1 for the areas classically involved with the commonly impaired lumbosacral roots and peripheral nerves; but note that areas are often incompletely involved, that sensation testing is subjective and thus often quite inaccurate unless the deficit is almost complete, and that there are often individual variations from the dermatomes shown.

Table 13-I   Strength Testing: Lower Extremity

All maneuvers are done against your resistance, and with the patient seated with legs dangling. Compare with the uninvolved side.

FLEXING THE HIP (raising the thigh)
(ILIOPSOAS; L1-2-3; femoral n.)

ADDUCTING THE HIPS (bringing the knees together)
(ADDUCTORS; L3-4; obturator n.)

ABDUCTING THE HIPS (spreading the knees apart)
(GLUTEI: L4-5-S1; superior gluteal n.)

EXTENDING THE KNEE (straightening the leg)
(QUADRICEPS; L2-3-4; femoral n.)

FLEXING THE KNEE (pulling heel under table)
(HAMSTRINGS; L4-5-S1; sciatic n.)

PLANTAR FLEXING THE FOOT (in neutral)
(GASTROCNEMIUS; L5-S1-2; tibial n.)

HOLDING THE PLANTAR FLEXED FOOT INVERTED
(POSTERIOR TIBIALIS; L5-S1; tibial n.)

HOLDING THE PLANTAR FLEXED FOOT EVERTED
(PERONEI; L5-S1; sup. peroneal n.)

DORSI FLEXING THE FOOT (in inversion)
(ANTERIOR TIBIALIS; L4; deep peroneal n.)

DORSI FLEXING THE GREAT TOE
(EXTENSOR HALLUCIS LONGUS; L4-5; deep peroneal n.)

Electrodiagnosis

When the diagnosis cannot be clearly defined by the history and examination described above, referral should be made for EMG-NCV. See Chatper 2.

X Ray

A lumbosacral spine X ray should be obtained in every patient with objective neurologic findings not clearly due to a peripheral problem. Referral for myelography is warranted in cases with severe or progressive deficit or if there is suspicion of a spinal cord lesion.

SYNDROMES

Meralgia Paresthetica (*Uncommon*)

Diagnosis

Numbness and often a burning sensation is felt in the lateral aspect of the thigh (Figure 13-1).

It must be differentiated from L4 and L3 root lesions.

If there is no evident cause, a diabetic mononeuropathy must be ruled out by checking blood sugars.

Pressure over the anterior iliac spines may cause tingling in the involved area of the thigh.

Pathophysiology

Compression of the lateral femoral cutaneous nerve occurs as it passes under the inguinal ligament just inferior to the anterior superior iliac spine.

This usually occurs in obese patients or those who wear constrictive girdles or heavy tool belts.

Treatment

Avoid compression by the responsible garment or belt by discontinuing its use, moving it or padding it.

Common Peroneal Nerve Compression (*Rare*)

Diagnosis

Foot drop is the most dramatic result. See Figure 13-1.

Tenderness is found at the fibular head.

Pathophysiology

Compression usually occurs where the nerve passes over the fibular head, either a result of trauma or of sitting "indian style" with the outsides of the knees on a hard floor.

Treatment

The nerve must be decompressed by treating the underlying cause; *referral* for this (as well as for bracing and rehabilitation for any foot drop) should be made.

Deep Peroneal Nerve Compression (*Uncommon*)

Diagnosis

Decreased sensation is found between the first and second toes, and sometimes mild weakness of the extensor of the great toe as well (Figure 13-1).

Tenderness occurs where the nerve becomes superficial in the anterior ankle.

This must be differentiated from common peroneal nerve injury and from the anterior compartment syndrome.

Pathophysiology

Compression at the anterior ankle occurs by shoes that are too tight or too tightly laced.

Treatment

Simply by *avoidance* of that compression.

Tarsal Tunnel Syndrome (*Rare*)

Diagnosis

A burning pain is felt in the sole of the foot, sometimes with radiation up the back of the leg.

Sensory loss can be found on the inferior heel or the sole of the foot. The specific area is variable depending on which branch of the nerve is most affected.

Tenderness is found over the nerve just behind the medial malleolus.

Pathophysiology

Compression of the posterior tibial nerve occurs as it passes under the lancinate ligament just posterior to the medial malleolus. The cause may be trauma, direct compression by a shoe, or excessive pronation of the foot.

Treatment

Correct the *underlying cause* as much as possible (especially excessive pronation).

Injection may be tried; technique is to prep, spray with ethyl chloride, and then to infiltrate $\frac{1}{2}$ cc each of steroid and local anesthetic without epinephrine just behind the medical malleolus. See Chapter 3 for further details and precautions.

If these measures fail or if weakness of the toe flexors at the proximal joints develops, the patient should be referred for electrodiagnostic confirmation and *surgical release*.

14

The Hip

Hip pain in *children* is covered in Chapter 19. Hip pain in *runners* is covered in Chapter 20. Further details on therapeutic suggestions made in this chapter are to be found in Chapter 3.

For the anatomy of the hip see Figure 14-1.

EVALUATION

Always consider the possibility of hip pain being *referred* from pelvic, intraabdominal or retroperitoneal disease. Although the differential diagnosis of these conditions is obviously beyond the scope of this book, they must be kept in mind and appropriate history, examination and tests done to rule them out, especially if the initial history and/or examination is not completely consistent with a musculoskeletal origin of the patient's pain.

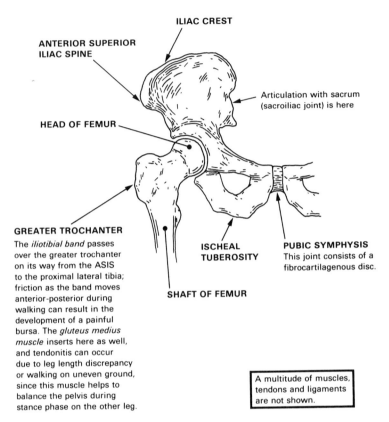

Fig. 14-1. Anatomy of the hip.

History

Ask about *duration* and *location* of pain. Pain from within the hip joint is usually felt either in the groin or the buttock, though it can be referred to the thigh or the knee.

What are *precipitating* and *relieving factors?*

Has there been any specific *trauma, strain,* or *overuse?*

Is there any *back pain* or *trauma,* any *radiation* of pain into the leg, or any *neurologic symptoms* in the thigh or leg? (If so, see Chapter 11 on the low back, or Chaper 13 on neurologic symptoms and diffuse pain of the lower extremity, as appropriate.)

Inquire about *previous* hip problems, diagnoses, or treatments.

Are any *other joints* involved? Is there a past history of arthritis or gout?

A complaint of ''snapping'' in the hip is most commonly due to the iliotibial band popping over the greater trochanter. (If so, the snapping will be palpable there as the hip is flexed and extended.)

Examination

Localize *tenderness* (Figure 14-2).
Observe the patient's *gait*.
Check *range of motion* (Figure 14-3). Pain in all directions of motion implies an intraarticular process.
Examine the *lumbosacral spine* and do the *straight leg raising test* to rule out referred pain from the low back. Also examine distal *neurovascular* status. The most common cause of the complaint of hip pain in the postadolescent presenescent patient is lumbosacral disease. See Chapter 11 if appropriate.
Measure *leg lengths* (from each anterior superior iliac spine to its medial malleolus) and also distances between the ASIS and the greater trochanters to detect hip joint shortening.
Do *abdominal, pelvic,* and *rectal* examination as appropriate.

X Ray

Although when X rays should be taken was discussed in Chapter 2, the criteria must be looser in the case of the hip because it is so inaccessible to palpation and is so often the site of metastatic disease. Any painful hip in which the diagnosis is not obviously both periarticular (e.g., trochanteric bursitis) and rapidly resolving should be X rayed.

HIP PAIN AND STRAIN

Trochanteric Bursitis (*Common*)

Diagnosis

Tenderness is found over the greater trochanter only (Figure 14-2).
Pain is elicited with external rotation and adduction only.

Pathophysiology

Several muscles attach at the greater trochanter; stresses at these insertions can lead to inflammation.

170

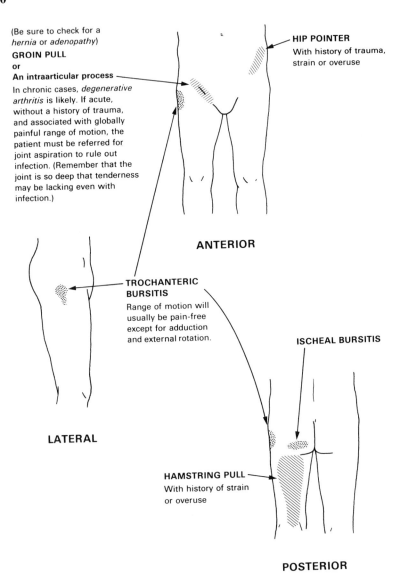

(Be sure to check for a *hernia* or *adenopathy*)
GROIN PULL
or
An intraarticular process
In chronic cases, *degenerative arthritis* is likely. If acute, without a history of trauma, and associated with globally painful range of motion, the patient must be referred for joint aspiration to rule out infection. (Remember that the joint is so deep that tenderness may be lacking even with infection.)

HIP POINTER
With history of trauma, strain or overuse

ANTERIOR

TROCHANTERIC BURSITIS
Range of motion will usually be pain-free except for adduction and external rotation.

ISCHEAL BURSITIS

LATERAL

HAMSTRING PULL
With history of strain or overuse

POSTERIOR

Fig. 14-2. Sites of tenderness around the hip.

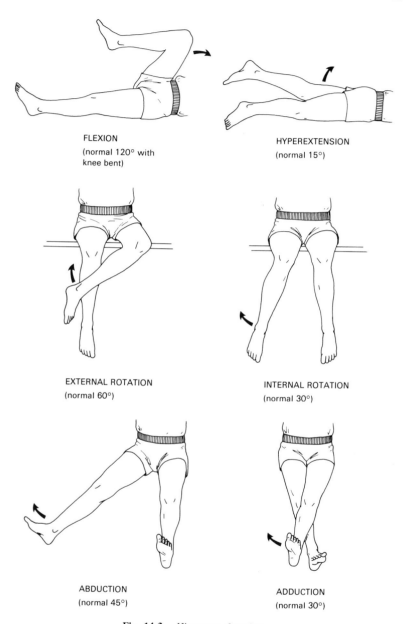

FLEXION
(normal 120° with
knee bent)

HYPEREXTENSION
(normal 15°)

EXTERNAL ROTATION
(normal 60°)

INTERNAL ROTATION
(normal 30°)

ABDUCTION
(normal 45°)

ADDUCTION
(normal 30°)

Fig. 14-3. Hip range of motion.

Or, a painful bursa can develop due to friction between the trochanter and the overlying iliotibial band.

Treatment

Try *rest, heat,* and *antiinflammatory medications.*
Steroid injection has a very good success rate and should be offered if needed. See Figure 14-4.

Ischeal Bursitis (Weaver's Bottom) (*Uncommon*)

Diagnosis

The patient gives a history of prolonged sitting, especially with the legs crossed and on a hard surface.
Tenderness is found only over the ischeal tuberosity (Figure 14-2).

Pathophysiology

Inflammation is due to prolonged pressure.

Treatment

Have the patient *avoid prolonged sitting,* especially with the legs crossed.

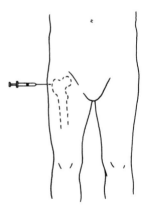

Fig. 14-4. Injection for trochanteric bursitis. Prep over the tender area and spray with ethyl chloride. Use a 23-gauge needle to inject a cc each of steroid and local anesthetic in several directions (without withdrawing from the skin) around the inflamed area. See pp. 23–27 for further details and precautions.

The use of a *soft pad* or a *ring seat* may be helpful.
Antiinflammatory medications can be used if the problem is acute.
Steroid injection has a good success rate. Use a 23-gauge needle to infiltrate a cc or so each of steroid and anesthetic around the tender area, after prep and ethyl chloride spray. See Chapter 3 for further details and precautions.

Groin Pull (*Common*)

Diagnosis

The patient reports acute or repetitive abduction strain.
Tenderness may be present over the inguinal ligament or the medial thigh.
Be sure to check that the patient does not have a hernia or lymphadenitis.

Pathophysiology

Strain is of the inguinal ligament and/or the adductor musculature.

Treatment

Use *ice* if acute.
Advise *rest*.
Antiinflammatory medications may hasten resolution if the pull is at all severe.
Once the episode has resolved, *stretching exercises* may help to prevent recurrence. See Patient Handout 20-2.

Hip Pointer (*Common only in athletes*)

Diagnosis

There often is a history of a direct blow to the iliac crest region. Alternatively, the patient may have been doing much running.
An X ray must be done to rule out an avulsion fracture of the anterior superior iliac spine or iliac crest. ·

Pathophysiology

A painful hematoma develops from direct trauma in the soft tissues that attach there.

174

Or this can be a traction injury to muscular origins.

Treatment

Use *ice* and *compression* acutely.
In teenagers, avoidance of strain and trauma is advisable for six to eight weeks. Older patients may return to activity as tolerated.

Hamstring Pull (*Common*)

Diagnosis

The patient gives a history of acute or repetitive strain.
Pain and tenderness may be located anywhere in the posterior muscles of the thigh.

Pathophysiology

Often the basic problem is quadriceps muscles relatively too strong for the opposing hamstring muscles.

Treatment

Use *ice* for the acute strain.
Rest is advisable.
Antiinflammatory medications may hasten resolution if the pull is at all severe.
Once the episode has resolved, *strengthening* and *stretching exercises* to prevent recurrence are advised. See Patient Handouts 20-2 and 20-3.

Degenerative Arthritis (Osteoarthritis) of the Hip (*Very Common*)

Diagnosis

Chronic pain is worse after weight-bearing activity, and is better with rest. The pain can be felt in the groin, the thigh, or the knee.
Found in older patients or those with previous significant hip disease or trauma.
There may be variable pain with range of motion.

Pathophysiology

See Chapter 22.

Treatment

See Chapter 22.

Weight loss is critical if the patient is obese.

The use of a *cane* can be quite helpful for some patients.

Referral for consideration of *injection* or *surgery* should be considered in severe cases not responding to conservative management.

15

The Knee

Knee problems in *children* and *adolescents* are covered in Chapter 19. Knee pain in *runners* is discussed in Chapter 20. Further details on therapeutic suggestions made in this chapter can be found in Chapter 3.

For the anatomy of the knee see Figure 15-1.

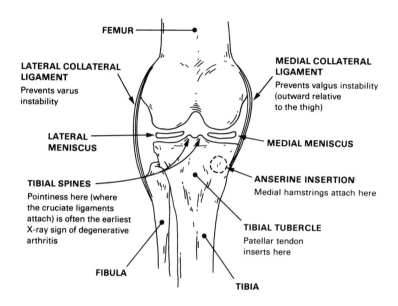

FEMUR

LATERAL COLLATERAL LIGAMENT
Prevents varus instability

MEDIAL COLLATERAL LIGAMENT
Prevents valgus instability (outward relative to the thigh)

LATERAL MENISCUS

MEDIAL MENISCUS

TIBIAL SPINES
Pointiness here (where the cruciate ligaments attach) is often the earliest X-ray sign of degenerative arthritis

ANSERINE INSERTION
Medial hamstrings attach here

TIBIAL TUBERCLE
Patellar tendon inserts here

FIBULA

TIBIA

**FRONT VIEW
(WITH PATELLA AND QUADRICEPS REMOVED)**

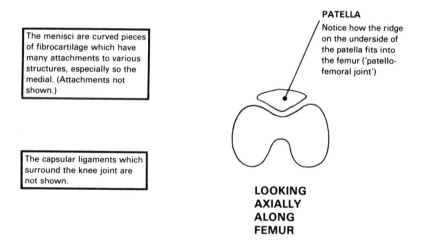

The menisci are curved pieces of fibrocartilage which have many attachments to various structures, especially so the medial. (Attachments not shown.)

PATELLA
Notice how the ridge on the underside of the patella fits into the femur ('patellofemoral joint')

The capsular ligaments which surround the knee joint are not shown.

LOOKING AXIALLY ALONG FEMUR

Fig. 15-1. Anatomy of the knee.

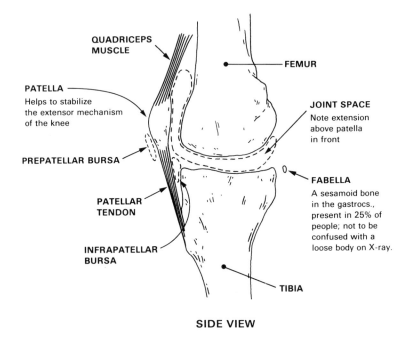

SIDE VIEW

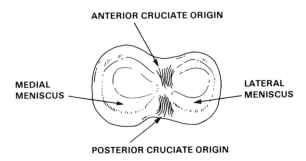

**LOOKING AT
TOP OF TIBIA**

Fig. 15-1. Continued.

HISTORY

Ask about the *duration* and *location* of pain.

What are *precipitating* and *relieving factors?*

Has there been any recent or previous *injury?* If so, what was the mechanism thereof?

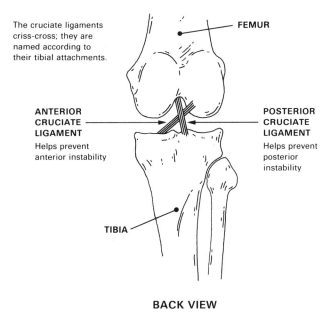

The cruciate ligaments criss-cross; they are named according to their tibial attachments.

FEMUR

ANTERIOR CRUCIATE LIGAMENT
Helps prevent anterior instability

POSTERIOR CRUCIATE LIGAMENT
Helps prevent posterior instability

TIBIA

BACK VIEW

Fig. 15-1. Continued.

If there was swelling after the injury, ask whether it occurred immediately (a sign of hemarthrosis, which in turn usually means a fracture or cruciate ligament tear), or a few hours later (which would be suggestive of an intraarticular lesion irritating the synovium, such as a meniscus tear).

If there was a noise at the time of injury, a pop is said to indicate an anterior cruciate tear, a crunch to indicate a meniscal tear, and a tearing sensation to be suggestive of a collateral ligament injury. These clues are not completely reliable.

Ask about any recurrent *swelling,* and whether it is associated with *redness* or *warmth.*

Does the knee ever *lock?* A locked knee is one that cannot be fully extended because of a physical block, not just because of pain or swelling; it is usually due to a torn meniscus. A patient seen after trauma with a locked knee should be seen by an orthopedist within 24 hours.

Does the knee *collapse?* Although it can occur to some extent in other conditions such as chondromalacia patellae, frequent collapse usually means:

Ligamentous instability

A torn meniscus

A loose body due to previous trauma or osteochondritis dissecans

A dislocating patella

Is there recurrent *clicking* which suggests a meniscus tear?

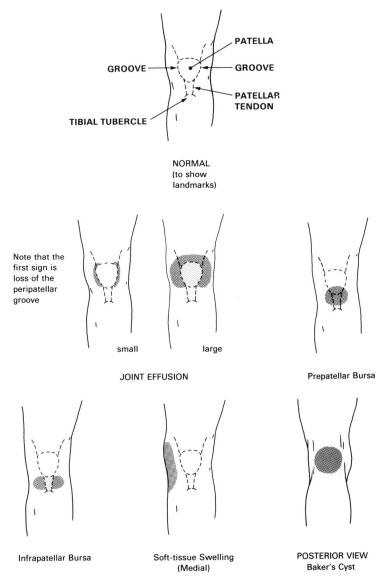

Fig. 15-2. Knee swelling.

Is there a *grating* sensation? (Common with chondromalacia patellae or degenerative arthritis).

Inquire about any *previous* knee problems, diagnoses, or surgery.

Find out how the problem *limits* the patient in his or her daily activities.

Are any *other joints* involved, or is there a history of arthritis or gout?

Always remember that knee pain may be *referred* from the hip or back. As a matter of fact, *the most common cause of knee pain in a child is pain referred from serious disease in the hip.*

EXAMINATION

Look for *quadriceps atrophy* by measuring the thigh circumference a measured distance proximal to the superior aspect of the patella. Compare to the unaffected leg. More than one cm of difference is significant and can develop in just a week of disuse.

Look for *effusion,* which will first be noticeable as a loss of the groove at the medial side of the patella. Joint effusion must be distinguished from swelling in the prepatellar or infrapatellar bursae, from a Baker's cyst, and from soft tissue swelling localized only to the traumatized area of the knee (Figure 15-2). An atraumatic effusion may mean infection, gout, or pseudogout and must be tapped (Figure 15-3).

Remember that the presence of *redness* or *warmth* may mean infection.

Test *range of motion.* Normal full flexion is 130°. Limitation of motion due to muscle spasm or inadequate relaxation must be distinguished from that due to effusion or to true physical limitation (as by a torn meniscus, for example). Note that a very swollen knee will be held at about 30° of flexion, since in this position the volume of the synovial space is greatest.

Look for locations of *tenderness.* See Figure 15-4.

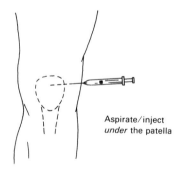

Aspirate/inject
under the patella

Fig. 15-3.   How to aspirate or inject a knee joint. Use sterile technique throughout. After proper preparation of the site, raise a wheal of local anesthesia under the superior-lateral aspect of the patella. Then inject some of the anesthetic in the direction shown. Now withdraw and use an 18-gauge or 16-gauge needle on a syringe of at least 20 cc capacity (a smaller syringe does not create enough vacuum on aspiration) in the direction shown. If steroid is to be injected, use a hemostat to hold the needle bevel and change syringes to one containing 2 to 6 cc of steroid. (Never inject steroid if the synovial fluid is cloudy or if there is other reason to suspect infection.) Steroid can be injected in the same way, without aspiration, when indicated, using a 22-gauge needle. Be sure to see pp. 13 and 23 for discussion of synovial fluid analysis and of precautions and risks of steroid instillation.

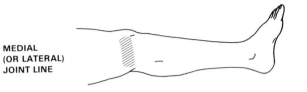

MEDIAL
(OR LATERAL)
JOINT LINE

The groove between
the femur and the
tibia is palpable.

Tenderness here can be in the ligament, in the meniscus
or can indicate degenerative arthritis.
The point of maximal tenderness will usually move posteriorly
a bit when the knee is flexed if a painful *MENISCUS* is responsible
(called Steinman's test.)
Tenderness may extend a bit above and below the joint line
if it arises from the *LIGAMENT.*
Joint-line tenderness due to *DEGENERATIVE ARTHRITIS*
is usually fairly diffuse.

PERIPATELLAR

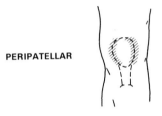

Tenderness here can be found in *CHONDROMALACIA
PATELLAE* or *DEGENERATIVE ARTHRITIS* of the patello-femoral
joint, but the patellar compression test is more specific.

**PATELLAR
TENDON**

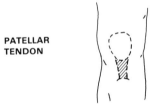

Tenderness anywhere here indicates *PATELLAR TENDONITIS*
(except at either end in a pre-teenager or adolescent: See *OSGOOD-
SCHLATTER'S DISEASE* in Chapter 19).

Fig. 15-4. Sites of tenderness around the knee.

Check *patello-femoral function* by performing the patellar compression and/or
patellar inhibition tests (Figure 15-5). Positive results imply the presence of
CHONDROMALACIA PATELLAE in younger patients, DEGENERATIVE
ARTHRITIS of the patello-femoral joint in older ones.

A positive patellar apprehension test (Figure 15-5) or excessive lateral

TIBIAL TUBERCLE

Tenderness (and sometimes swelling) here in a teenager implies *OSGOOD-SCHLATTER'S DISEASE* (covered in Chapter 19).

The bony prominence where the anserine tendons insert is easily palpable inferior to the joint line on the antero-medial tibia.

Tenderness here indicates *ANSERINE TENDONITIS.*

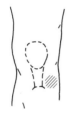

GERDY'S TUBERCLE
on the anterolateral tibia

Tenderness here indicates *ILIOTIBIAL BAND TENDONITIS* (covered in Chapter 20).

Fig. 15-4. Continued.

mobility of the patella is suggestive of RECURRENT PATELLAR DISLOCA-TION or subluxation (discussed in Chapter 19), if the history is compatible.

Check the *stability* of the *collateral ligaments* by stabilizing the femur with the knee flexed at 30° (because in full extension other structures are tight), and then exerting a valgus (to test the medial collateral) and then a varus (lateral collateral) stress (Figure 15-6). Getting complete muscular relaxation is the trick. Always compare to the other knee. In acute sprains where you cannot be completely certain that there is no laxity, the knee should be immobilized in a posterior splint and seen by an orthopedist within 72 hours.

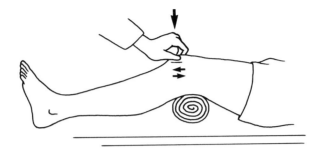

Patellar Compression
Press the patella down into the femur and move it proximally and distally. Pain and/or crepitus is a positive test.

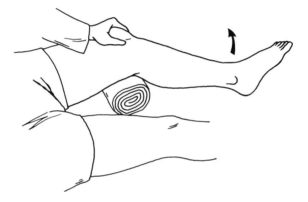

Patellar Inhibition
With the knee flexed about 20° (over your other hand or a rolled towel), put your thumb and forefinger around the top of the patella and ask the patient to straighten his leg. If he starts to, and then stops because of pain, the test is positive.

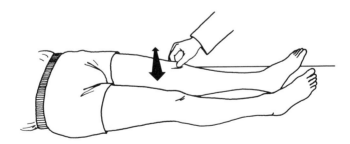

Patellar Apprehension
Have the patient relax and then start to move his patella laterally. If he resists, the test is positive.

Fig. 15-5. Patello-femoral tests.

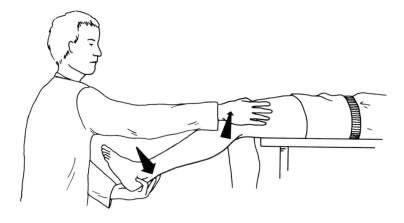

MEDIAL COLLATERAL LIGAMENT

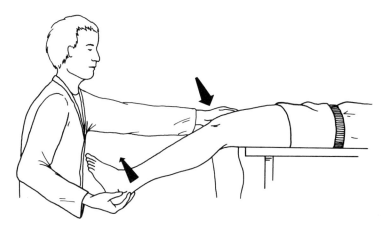

LATERAL COLLATERAL LIGAMENT

Good muscular relaxation is essential
Don't allow the thigh or leg to rotate while performing the test
Always compare to the uninvolved knee

Fig. 15-6. Checking collateral ligament stability.

Check the integrity of the *cruciate ligaments* by performing the drawer tests (Figure 15-7). In chronic instability cases, perform the anterior drawer test with the foot in external rotation as well as in internal rotation and neutral to pick up partial tears of the posterior capsule. The *pivot shift* test (Figure 15-8) may be used; it is somewhat more sensitive than the anterior drawer test for detecting chronic anterior cruciate laxity.

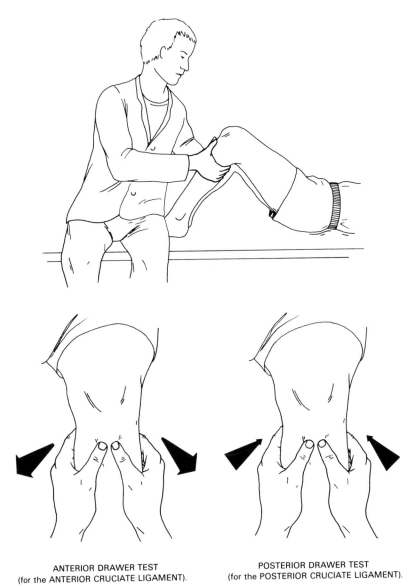

ANTERIOR DRAWER TEST
(for the ANTERIOR CRUCIATE LIGAMENT).

POSTERIOR DRAWER TEST
(for the POSTERIOR CRUCIATE LIGAMENT).

Fig. 15-7. The drawer tests. Getting complete muscular relaxation is necessary for a good exam. Always compare to the uninvolved leg. Asymmetric laxity implies a torn cruciate ligament.

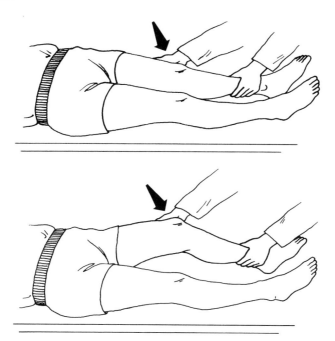

Fig. 15-8. The pivot shift test. A palpable or visible pop forward of the tibia during early flexion with a valgus stress implies anterior cruciate ligament laxity.

Test for meniscal tears by performing *McMurray's test* (Figure 15-9), which however is often negative even with a tear. If the patient can do a good *deep knee bend* or *duck-walk* (Figure 15-10) without difficulty, medial meniscal problems are unlikely.

Apley's test (Figure 15-11) can help differentiate ligamentous from meniscal problems.

Having the patient *hop* on each leg is a good screening test for significant knee disability.

KNEE PAIN

Evaluation

See the discussion of history and examination above.

Chondromalacia Patellae (*Very Common*)

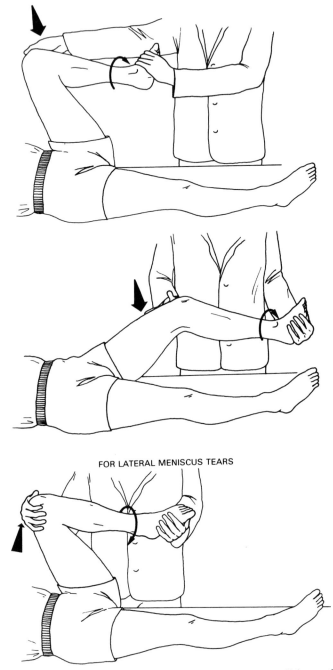

FOR MEDIAL MENISCUS TEARS

FOR LATERAL MENISCUS TEARS

Fig. 15-9. Murray's test. The top two drawings on this page illustrate medial tears; the bottom one here and the top one on the next page illustrate lateral tears.

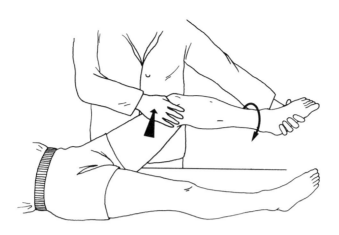

Fig. 15-9. Continued.

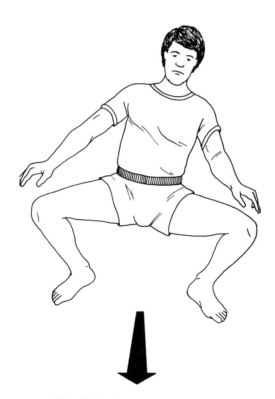

Fig. 15-10. The duck-walk maneuver.

Distraction Test (painful with tender ligaments) (external rotation: Laterial ligaments)

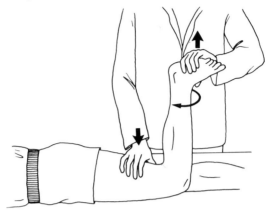

Compression Test (painful with torn meniscus, sometimes with a click as the leg is extended (external rotation: Medial meniscus)

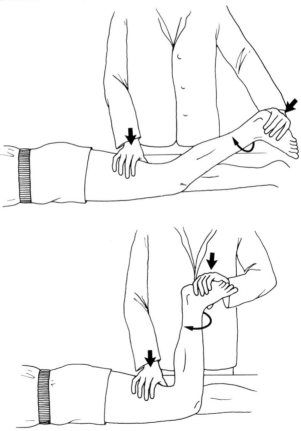

Fig. 15-11. Apley's test. Good muscular relaxation is necessary in both variations. The top drawings on these two pages illustrate a distraction test, the other four a compression test.

Distraction Test (Continued) (internal rotation: Medial ligaments)

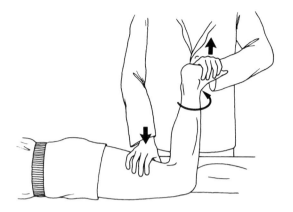

Compression Test (Continued) (internal rotation: Lateral meniscus)

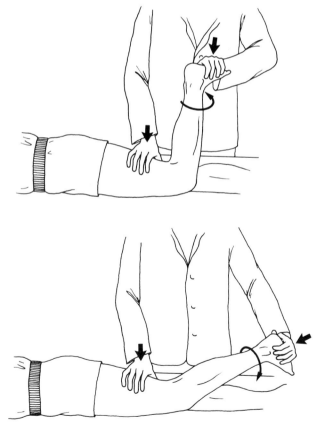

Fig. 15-11. Continued.

Table 15-I   Examination of the Knee

With the patient lying supine:

1. Measure each thigh circumference 6 cm proximal to the superior pole of the patella.
2. Look for joint effusion or other swelling.
3. Check passive range of motion.
4. Look for sites of tenderness.
5. Check medial and lateral stability with the knee flexed 30° and then with it straight.
6. Do the anterior and posterior drawer tests.
7. Perform the patellar compression and apprehension maneuvers.
8. Do McMurray's test.

Have the patient roll over to the prone position:

9. Look for tenderness or swelling in the popliteal fossa.
10. Do Apley's distraction and compression tests.

Have the patient stand up:

11. Ask the patient to do a deep knee bend on each leg.
12. Have the patient duck-walk.
13. Have the patient hop on each leg.

With some practice the entire examination will take no more than a few minutes. Of course, additional examination should be done if indicated.

Diagnosis

The patient complains of an aching pain worse after activity, especially after going up and down stairs or hills.

There may also be a grating sensation.

Sometimes the patient reports occasional collapse.

There is stiffness after prolonged immobility (the "movie sign").

Most commonly found in teenagers and young adults, especially females.

Examination reveals tenderness along and under the patellar border and a positive patellar compression test. (See Figure 15-5).

Note: Chondromalacia is a very common asymptomatic finding, and its presence on examination does not necessarily mean that it is the etiology of the patient's particular complaint.

Pathophysiology

Symptoms are due to irregularity and roughness of the cartilage that lines the underside of the patella.

This may develop because of a congenitally poor fit between the underside of the patella and the femoral groove in which it moves (see Figure 15-1), or to misalignment of the pull of the quadriceps, or to previous trauma causing a chip off the cartilage on the underside of the patella. Sometimes excessive foot pronation can be responsible, by leading to internal torsion of the leg and thus increased compressive force at the medial side of the patello-femoral joint.

Treatment

The patient should *rest* only as needed to reach a tolerable level of symptoms. *Aspirin* and *heat* may be used as needed.

A program of *quadriceps strengthening exercises* (Patient Handout 15-1) gives added stability to the patella and realigns the pull of the quadriceps. It is often extremely helpful.

Consider treatment to decrease excessive foot pronation (described in Chapter 20), especially if mainly the medial side of the patella is involved.

Surgery (to shave smooth the offending cartilage or realign the pull of the quadriceps muscle) should be considered a very last resort, since the natural history of the condition is improvement with time.

Medial Meniscus Tears (Torn Cartilage) (*Common*)

Diagnosis

The patient often gives a history of twisting the knee with the foot planted, leading to sudden pain in the knee (with or without an associated crunching sensation), followed by swelling several hours later.

Or there may be a history of chronic or recurrent knee pain and recurrent swelling. Note that history of the inciting trauma is often completely lacking in these chronic cases.

The knee may recurrently collapse.

Medial joint line tenderness can be found on examination.

Often there is also pain on approach to full flexion.

The patient is usually unable to do a good deep knee bend on the involved knee, and duck walking may lead to pain and/or clicking (Figure 15-10).

McMurray's and Apley's compression tests (Figures 15-9 and 15-11) are fairly diagnostic if positive but are often negative.

Uncommonly a firm lump may be felt over the joint line (the so-called *meniscal cyst*).

It is very common for an arthrogram or arthroscopy to be necessary to make or confirm the diagnosis, especially in chronic cases.

A fold of synovial membrane (*plica*) can often mimic a meniscal tear, the correct diagnosis only being made on arthroscopy.

And often, lesions of the articular cartilage of the medial and lateral compartments can be indistinguishable by history and examination from meniscus pathology.

Pathophysiology

The meniscus is caught between the femur and tibia and torn, or torn because of stresses on its various connections.

Older patients in whom the menisci have degenerated to some extent require only very minor trauma to tear them ("degenerative meniscal tear").

Treatment

Acute Tears:

Immobilize the knee for one to several weeks or merely decrease weight-bearing, depending on severity.

The patient should be instructed on *protective* measures (not walking on uneven ground or on high heels, no activities which require sudden starts and stops, etc.).

Quadriceps strengthening exercises (Patient Handout 15-2) should be used to prevent atrophy.

Heat and *analgesics* may be prescribed for symptomatic relief.

Meniscal tears sometimes heal over six weeks or so, especially those of the more fibrous outer portion which has a better blood supply, and in many cases no surgery will be necessary.

However, a locked knee (one in which full extension is impossible) must be seen by an orthopedist within 24 hours, and severe tears (associated with much swelling and collapsing) are also best handled by the specialist.

In addition, any case in which an associated ligamentous tear cannot be definitely ruled out should be referred.

Patients not presenting with the acute episode:

There is no good evidence that degenerative changes in the knee occur any faster with a torn meniscus left in place than they do postsurgery. Therefore:

If the pain is severe enough or the disability great enough that the patient would consider surgery, or if the patient's knee is frequently collapsing, he should be *referred* to the orthopedist for definitive diagnosis and treatment.

But if those conditions are not present, treat with prn *rest, heat,* and *analgesics/antiinflammatories,* with option for referral later.

Lateral Meniscus Tears (*Uncommon*)

Diagnosis

Analogous to medial meniscus tears, discussed earlier.

Pathophysiology

These are much less common than tears of the medial meniscus, due to the greater mobility of the lateral.

A developmental "discoid" meniscus can mimic a tear, especially in young girls.

Treatment

Analogous to medial meniscus tears, discussed earlier.

Patellar Tendonitis (Jumper's Knee) (*Common*)

Diagnosis

Tenderness is found over the patellar tendon or its insertion at the tibial tubercle (Figure 15-4). (But if the patient is a teenager, see OSGOOD-SCHLATTER'S DISEASE in Chapter 19.)

Pathophysiology

Inflammation is due to excessive stress, usually due to sudden contraction of the quadriceps (necessary at the moment of heel strike in jumpers or runners to prevent knee collapse, since the knee is somewhat flexed at that moment).

Treatment

Rest by decreasing or eliminating the responsible activity.
Use *aspirin* or a course of *NSAIs*.
Heat may be helpful.
Injection is contraindicated due to the great stresses at this point and the risk of subsequent rupture.

Anserine Tendonitis (*Common*)

Diagnosis

Tenderness is at the anserine tendon insertion (Figure 15-4).

Pathophysiology

Excessive strain there leads to a tendonitis, or to inflammation of a bursa that develops between tendon and bone proximal to the insertion.

Treatment

Use *rest, heat,* and *antiinflammatories.*
Steroid injection has a good response rate. See Figure 15-12.

Degenerative Arthritis of the Knee (*Very Common*)

Diagnosis

Chronic pain is worse after weight-bearing activity, better with rest.
Found in older patients and those with previous significant trauma.
Tenderness may be in the medial joint line (medial compartment), lateral joint line (lateral compartment), along the patellar borders with a positive patellar compression test (patello-femoral compartment), or any combination thereof.
Occasionally there is some swelling.
There is a variable amount of pain with range of motion.
The X ray may be normal early (when changes are limited to cartilage), but some findings are usually present. (But remember that the presence of degenerative changes on X ray does not prove that any particular symptoms are due to those changes.)

Pathophysiology

See Chapter 22.

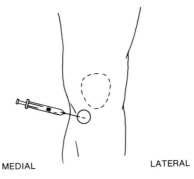

MEDIAL   LATERAL

Fig. 15-12. Injecting the Anserine Tendon Insertion. After prepping the area and spraying with ethyl chloride, use a 23 or 25-gauge needle to infiltrate a cc of steroid and a cc of local anesthetic around the tender area. See pp. 23–27 for further details and precautions.

Treatment

See Chapter 22. *Weight loss* is critical if the patient is obese.
Technique of *injection* is shown in Figure 15-3.
Referral for arthroscopic or open joint debridement, for osteotomy, or for total joint replacement should be considered in severe cases not responding to conservative management.

KNEE COLLAPSE

Evaluation

See the opening of this chapter. Some specific points follow.
Collapse only with certain sudden changes in direction (as in racket sports or dance, for instance) hints at CHRONIC LIGAMENTOUS INSTABILITY.
If there is swelling after the episodes of collapse, consider a chronic MENISCUS TEAR, a loose body from previous trauma or OSTEOCHONDRITIS DISSECANS, or a SUBLUXING PATELLA (the latter is discussed in Chapter 19).
Ask if the kneecap is displaced to the outside with the episode (implying a SUBLUXING PATELLA).
Collapse can occur with CHONDROMALACIA PATELLAE but is usually not frequent.
X ray may reveal a loose body if it is calcified (not to be confused with the fabella: see Figure 15-1).
A "tunnel view" should be requested to look for OSTEOCHONDRITIS DISSECANS and a "sunrise view" to look at the fit of the patella in its groove.

Chronic Ligamentous Instability (*Uncommon*)

Diagnosis

The patient will report that the knee gives out with sudden stops, starts or direction changes.
A history of previous sprain may sometimes be lacking.
Ligamentous laxity is found on examination (usually the medial collateral ligament, the anterior cruciate ligament, or the posterior capsule).

Pathophysiology

A result of previous injury.

Treatment

Knee *strengthening exercises* (Patient Handout 15-2) should be prescribed.
Use of a *brace* may be helpful.
Referral to an orthopedist for evaluation and consideration of *surgery* may be necessary.

Osteochondritis Dissecans (*Uncommon*)

Diagnosis

Usually found in teenagers and young adults.
There is sometimes pain or tenderness under the patella.
Often the patient gives a history of knee collapse.
An area of osteochondritis may be visible on a "tunnel view" X ray; a loose body may be visible if it is calcified.
Tomograms or referral for arthrography or arthroscopy may be necessary to make the diagnosis.

Pathophysiology

A piece of articular cartilage with or without a piece of attached bone breaks off the femoral surface (most commonly the lateral border of the medial condyle) and becomes a loose body in the joint.
This may be due to an anomaly of ossification, localized ischemia, or injury.

Treatment

If symptoms of pain or collapse are severe enough, *refer* the patient to an orthopedist for surgery.
Otherwise, treat *symptomatically* only.

KNEE SWELLING

Evaluation

Swelling of the prepatellar bursa or the infrapatellar bursa must be differentiated from swelling of the knee joint itself (Figure 15-2). The latter implies an intraarticular process; if acute and atraumatic, leading possibilities are infection, gout, and pseudogout. (The last-mentioned is especially common in the knee joint; a clue to its presence is an X ray showing calcium deposits in the menisci or articular cartilage. Pseudogout is discussed in Chapter 24.) Swelling in the posterior aspect of the joint is the so-called BAKER'S CYST.

Prepatellar Bursitis (Housemaid's Knee) (*Common*)

Diagnosis

A painless or slightly painful swelling is localized over the inferior patella and the patellar tendon (Figure 15-2).

If there is redness, warmth, significant tenderness, or a nearby break in the skin, it *must be tapped* to rule out infection.

Pathophysiology

Fluid accumulates in the inflamed bursa usually from prolonged kneeling.

Treatment

The patient should *avoid kneeling;* a *pad* may be used to prevent recurrence if kneeling is unavoidable.

Heat may help.

Drainage may be offered; the fluid usually reaccumulates quickly unless some steroid is instilled after drainage, but even this often does not work. The method is simply to enter the center of the swelling with a #18 needle (with a large syringe) after sterile preparation and ethyl chloride spray; never inject steroid if the fluid drained appears cloudy or purulent. Remember, the bursa must be aspirated if the signs of infection described above are present.

Surgery may be done to remove a chronically or recurrently inflamed bursa.

Infrapatellar Bursitis (*Uncommon*)

Diagnosis

A painless or slightly painful swelling is found as shown in Figure 15-2.

Pathophysiology

The bursa becomes inflamed from overuse (flexion/extension) of the knee.

Treatment

Rest, heat, and *antiinflammatories* may be useful.

Baker's Cyst (*Uncommon*)

Diagnosis

A painless or slightly painful swelling occurs in the popliteal space (Figure 15-2).

Be sure it is not pulsatile or associated with a bruit, since a popliteal artery aneurysm can be misdiagnosed; and be sure it is cystic, because soft tissue tumors can present in this location.

Pathophysiology

This may actually be a bursa associated with the semimembranosus or gastrocnemius, which may or may not be in communication with the joint space; or it can be a posterior extension of the synovial space of the knee.

Swelling of the latter type is associated with intraarticular knee problems.

The cyst can occasionally rupture, leading to a swollen inflamed calf which may mimic deep vein thrombosis.

Treatment

Treatment of the *underlying knee problem* is the best treatment of the cyst.

Surgical excision should be resorted to only if the above is not successful and the cyst is sufficiently large and symptomatic.

KNEE TRAUMA

Evaluation

Always examine distal neurovascular status. History and examination are discussed in the beginning of this chapter. Always X ray. *Note:* If joint swelling (as opposed to localized soft tissue swelling) is immediate after injury, the patient probably has a *hemarthrosis,* indicating a fracture (even if not visible on X ray) or cruciate ligament tear, and should be referred immediately to an orthopedist. Similarly, all fractures should be referred at once. MENISCAL TEARS are discussed above.

Knee Sprains (*Common*)

Diagnosis

A twisting injury occurs.

If there is *noticeable laxity* in any direction (see Figures 15-6 and 15-7 for method of examination), or if *examination is inadequate* because of swelling, incomplete muscle relaxation by the patient or any other reason (i.e., ligamentous laxity cannot be completely ruled out), the knee should be *immobilized* in a posterior splint and *be seen by an orthopedist within 72 hours*. (This is because primary repair is the treatment of choice, and is best done in the first week or so after injury.)

X ray will be normal. (An avulsion chip indicates a tear and must be handled as noted above.)

Treatment (if laxity is definitely ruled out)

Prescribe *immobilization* and *non-weight-bearing,* with the use of crutches.
Elevation and elastic wrap for *compression* are advisable.
Ice should be applied during the first 48–72 hours.
Analgesia should be prescribed as needed.
Progressive weight-bearing and *knee motion* are resumed as tolerated as the days go by. If immobilization is for more than a few days, start *quadriceps exercises* to prevent atrophy. See Patient Handout 15-2.

Acute Patellar Dislocation (*Uncommon*)

Diagnosis

This may be a result of a direct blow or indirect strain.
The patella invariably dislocates laterally.

Treatment

These patients should be *referred* to an orthopedist immediately even if the patella has spontaneously relocated, since associated fracture or soft-tissue injury is common, and the knee is usually casted for four to six weeks to help prevent recurrence.

Patient Handout 15-1

Chondromalacia

Your doctor has diagnosed your knee problem as *chondromalacia patellae* (sometimes called the *patello-femoral syndrome* or *patello-femoral dysfunction*). You are not alone. Except for injuries, this is by far the most common cause of knee problems in young people.

What is it?

Your kneecap (or *patella*) is not flat on the underside, as a lot of people think. It actually has a ridge there. This ridge is supposed to slide up and down in a groove on the thigh bone (femur) as your knee bends during walking and running.

Now, every joint in the body has a lining of cartilage (gristle) on the bones where they meet, and the joint between the patella and the femur is no exception. If this cartilage becomes worn and roughened, you may feel an aching pain behind your kneecap after walking, especially after going up and down stairs or hills; you may even feel a grating sensation. The knee will feel stiff after you sit with it bent for a prolonged time. Occasionally a rough spot may catch and the knee will give out. The knee can even swell sometimes.

Why does the cartilage become rough?

A variety of causes are possible. The most common is weak thigh muscles; they then pull the kneecap crookedly, and the surface becomes worn. Or sometimes people are born with abnormalities that predispose to the same thing. Or poor foot posture can lead to strain here. Or occasionally injury to the kneecap starts the whole problem by chipping the cartilage.

What can I do about it?

Keep in mind that this condition won't lead to serious problems even if you ignore it altogether; as a matter of fact, the worn cartilage tends to smooth out as time goes on, and symptoms tend to go away. But if you want to do something, exercises to strengthen your thigh muscles are your best bet. You can also use heat on your knee and take aspirin when it is hurting.

Quad Sets

While standing, *slowly* tighten your thigh muscles (pulling your kneecap upward) and keep them tight for a count of ten; then slowly relax. Do this at least ten times an hour whenever you are standing. After a while you will be doing it automatically, without even having to think about it.

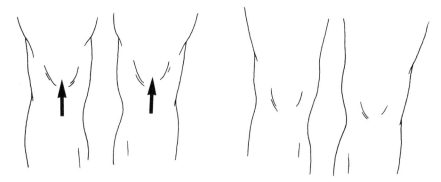

Straight Leg Raises

While lying on your back on the bed or floor, *slowly* raise your leg about a foot off the surface, keeping your knee straight. Hold it for a count of five and then slowly put it down. Do 10 or 15 repetitions twice a day. As you get stronger, you may even want to put weights around your ankles (books in a handbag do quite nicely). Athletes can lift about a quarter of their body weight during these exercises.

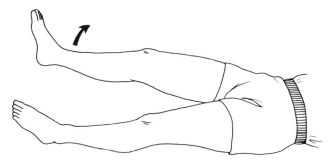

Patient Handout 15-2

Knee Exercises

Do the quadriceps exercises twice a day. Do the other exercises only if your doctor recommends.

QUADRICEPS EXERCISES

Quad Sets

While standing, slowly tighten your thigh muscles (pulling your kneecap upward) and keep them tight for a count of ten; then slowly relax. Do this at least ten times an hour whenever you are standing. After a while you will be doing it automatically, without even having to think about it.

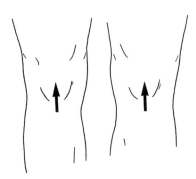

Straight Leg Raises

While lying on your back on the bed or floor, slowly raise your leg about a foot off the surface, keeping your knee straight. Hold it for a count of five and then slowly put it down. Do 10 to 15 repetitions twice a day. As you get stronger, you may even want to put weights around your ankle (books in a handbag do quite nicely.) Athletes can lift about a quarter of their body weight during these exercises.

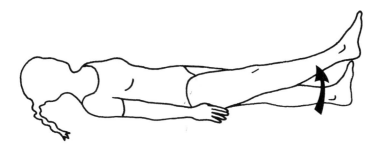

HAMSTRING EXERCISES

While lying on your stomach, raise your foot (by bending your knee) slowly into the air and hold for a slow count of five. Then slowly put it back down. Do ten to fifteen repetitions twice a day. As you get stronger, add weights around your ankle; aim for fifteen percent of your body weight.

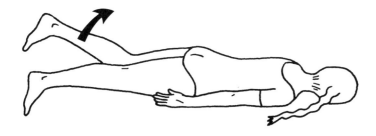

ABDUCTOR EXERCISES

While lying on your side (with the affected leg up) raise your leg (at the hip) about a foot into the air and hold for a slow count of five. Then slowly put it back down. Do ten to fifteen repetitions twice a day.

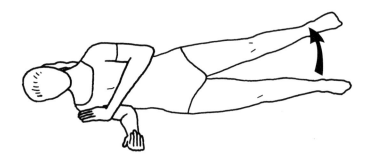

16

The Lower Leg

    Neurologic symptoms and diffuse pain of the lower leg are discussed in Chapter 13. Because *shin pain* is almost invariably a complaint of athletes, discussion is deferred to Chapter 20. *Cramps* are covered in Chapter 21.

CALF PAIN

Evaluation

    Calf pain can of course be due to *Deep Vein Thrombosis,* a potentially life-threatening condition. The diagnosis of this condition is beyond the scope of this book (and in fact is quite difficult to make without performing invasive tests). Its possibility must never be forgotten, and history, examination and tests appropriately performed to rule it out before one of the conditions discussed in this section is diagnosed.

    A ruptured BAKER'S CYST is another cause of acute calf pain and swelling that must be considered.

    Calf pain that comes on with activity and is relieved with rest in a fairly reproducible pattern is suggestive of *Claudication,* and appropriate examination and tests must be performed to investigate this possibility.

    Ask about any specific *trauma, strain* or *overuse.* A *snap* felt in the calf followed by pain and occuring with a sudden direction change is suggestive of

PLANTARIS RUPTURE, while a less sudden and less dramatic onset may occur with a CALF PULL.

Always consider the possibility of pain in the calf being *referred* from the low back. See Chapter 13 if appropriate.

Plantaris Rupture (Tennis Leg) (*Uncommon*)

Diagnosis

A sudden *snap* is felt in the calf, usually upon a sudden direction change, and is followed by pain and sometimes some swelling and ecchymosis.

Be sure that the patient has *full strength* in ankle plantar flexion (i.e., that you are not dealing with a torn achilles tendon, a much more serious matter). There should be no palpable gap in the achilles tendon or the calf.

Pathophysiology

This is a rupture of the plantaris tendon, a long thin structure that runs external to the soleus muscle all the way from its small muscle belly in the back of the knee to the posterior heel. A similar syndrome can occur with a partial rupture of the gastrocnemius.

Treatment

Elevation and *ice* packs acutely and *decreased activity* for a couple of weeks are all that is needed. No significant disability follows this injury.

Calf Pull (*Common*)

Diagnosis

The patient gives a history of acute or repetitive strain.
Pain and tenderness can be located anywhere in the calf.

Treatment

Use *ice* in the acute situation.
Advise *rest*.
Antiinflammatory medications may hasten resolution if the pull is severe.
A *heel lift* may be of temporary symptomatic benefit.
To prevent recurrence, *strengthening* and *stretching exercises* should be prescribed. See Patient Handouts 20-2 and 20-3.

17

The Ankle

Problems around the *achilles tendon* and *heel* are covered in Chapter 18. Ankle and heel cord pain in *children* are covered in Chapter 19. Ankle pain in *runners* is discussed in Chapter 20. For further details on therapeutic suggestions made in this chapter, see Chapter 3.

For the anatomy of the ankle see Figure 17-1.

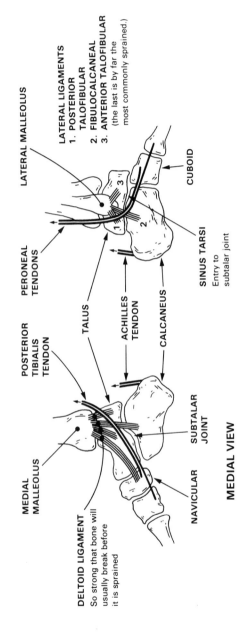

MEDIAL VIEW

MEDIAL MALLEOLUS

POSTERIOR TIBIALIS TENDON

TALUS

ACHILLES TENDON

CALCANEUS

NAVICULAR

SUBTALAR JOINT

DELTOID LIGAMENT
So strong that bone will usually break before it is sprained

LATERAL VIEW

LATERAL MALLEOLUS

PERONEAL TENDONS

CUBOID

SINUS TARSI
Entry to subtalar joint

LATERAL LIGAMENTS
1. POSTERIOR TALOFIBULAR
2. FIBULOCALCANEAL
3. ANTERIOR TALOFIBULAR
(the last is by far the most commonly sprained.)

Anterior tibial and extensor tendons which cross the front of the ankle joint are not shown; nor are flexor tendons nor a multitude of minor ligaments.

Fig. 17-1. Anatomy of the ankle.

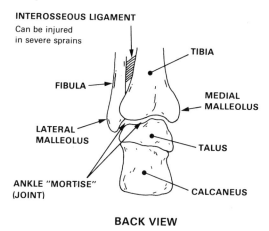

INTEROSSEOUS LIGAMENT
Can be injured
in severe sprains

TIBIA

FIBULA

MEDIAL
MALLEOLUS

LATERAL
MALLEOLUS

TALUS

ANKLE "MORTISE"
(JOINT)

CALCANEUS

BACK VIEW

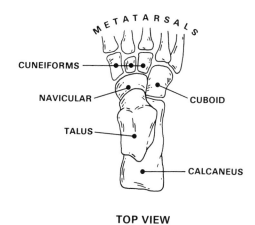

METATARSALS

CUNEIFORMS

NAVICULAR

CUBOID

TALUS

CALCANEUS

TOP VIEW

Fig. 17-1. Continued.

ANKLE PAIN

History

Has there been any specific *trauma, strain* or *overuse?*
What is the *duration* and the *location* of pain?
Does the ankle *twist recurrently?*
What are *precipitating* and *relieving* factors?

Ask about any *previous* ankle problems, diagnoses or treatments.
Are any *other joints* involved? Is there any history of arthritis or gout?

Examination

Localize *tenderness* (Figure 17-2).
Any generalized *joint swelling* in the absence of trauma implies an intraarticular process (which if acute must be tapped to rule out infection: see Figure 17-3). This must not be confused with the much more common finding of edema in the soft tissues secondary to venous insufficiency, congestive heart failure and other conditions. See Figure 17-4.
Any *redness* or *warmth* of course indicates an acute inflammatory process.
Test *range of motion*. Normal is 20° dorsiflexion, 45° plantar flexion, 30° inversion, and 20° eversion. Note that the last two numbers include much motion in the subtalar joint.
Test sensation, strength, pulses and distal capillary filling.

Posterior Tibialis Tendonitis (*Common*)

Diagnosis

Tenderness is found at the tendon's insertion on the tarsal navicular or first cuneiform bone and/or along its course behind the medial malleolus (Figure 17-2).

Pathophysiology

Excessive stretch is due to excessive foot pronation.
Sometimes the tendon inserts on an *accessory navicular bone*, and the pseudoarthrosis between the accessory and the true navicular becomes painful.
The problem may rarely be a residual inflammation after trauma, or be due to pressure from ill-fitting shoes.

Treatment

All of the measures discussed under CHRONIC FOOT STRAIN in Chapter 18 should be employed.
The sore area should be *padded* to reduce pressure.
In addition, a course of oral *antiinflammatory medication* may be helpful.
In prolonged or severe cases, *steroid injection* may be offered. Success rate is good if abnormal mechanics are corrected as discussed above. Technique is

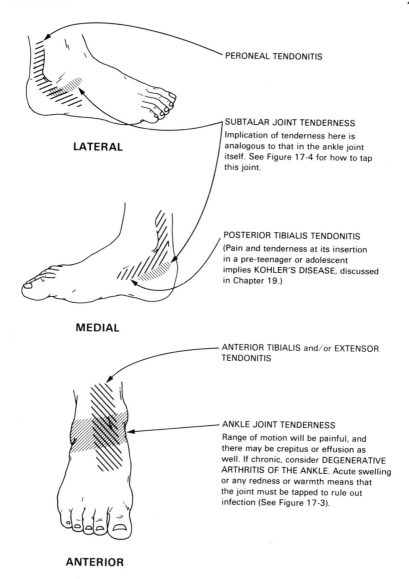

PERONEAL TENDONITIS

LATERAL

SUBTALAR JOINT TENDERNESS
Implication of tenderness here is
analogous to that in the ankle joint
itself. See Figure 17-4 for how to tap
this joint.

POSTERIOR TIBIALIS TENDONITIS
(Pain and tenderness at its insertion
in a pre-teenager or adolescent
implies KOHLER'S DISEASE, discussed
in Chapter 19.)

MEDIAL

ANTERIOR TIBIALIS and/or EXTENSOR
TENDONITIS

ANKLE JOINT TENDERNESS
Range of motion will be painful, and
there may be crepitus or effusion as
well. If chronic, consider DEGENERATIVE
ARTHRITIS OF THE ANKLE. Acute swelling
or any redness or warmth means that
the joint must be tapped to rule out
infection (See Figure 17-3).

ANTERIOR

Fig. 17-2. Sites of tenderness: the ankle.

simply to inject (after sterile prep and ethyl chloride spray) a cc or so of steroid and a cc or so of local anesthetic around the painful area (*never* into the tendon) using a 25-gauge needle. See Chapter 3 for further details and precautions before injecting.

Surgery may be resorted to in the presence of an accessory navicular.

Fig. 17-3. How to aspirate or inject an ankle joint. Find the large extensor hallucis longus tendon in the front of the ankle (have the patient dorsiflex the big toe to make it stand out), and then find the joint line (between the tibia and the talus) just medial to it. Prepare the site and use sterile technique throughout. Inject some anesthetic in the direction shown, then withdraw and insert a 20-gauge or 18-gauge needle on a 20-cc syringe in the same direction, and aspirate. If steroid is to be injected, use a hemostat to hold the needle bevel and change syringes to one containing 3 to 6 cc of steroid. (Never inject steroid if the synovial fluid is cloudy or if there is any other reason to suspect infection.) Steroid can be injected in the same way, without aspiration, using a 22-gauge needle. Be sure to see pp. 13 and 23 for discussion of synovial fluid analysis and of precautions and of risks of steroid instillation.

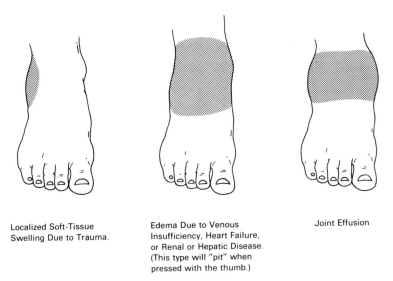

Localized Soft-Tissue
Swelling Due to Trauma.

Edema Due to Venous
Insufficiency, Heart Failure,
or Renal or Hepatic Disease.
(This type will "pit" when
pressed with the thumb.)

Joint Effusion

Fig. 17-4. Swelling around the ankle.

Peroneal Tendonitis (*Uncommon*)

Diagnosis

Tenderness is at the peroneus brevis tendon's insertion at the base of the fifth metatarsal and/or along the course of the peroneal tendons behind the lateral malleolus (Figure 17-2).

Pathophysiology

This usually follows a typical inversion sprain of the ankle.

Rarely it can be a result of excessive stretch in patients with a high arched foot and too much supination, or due to local pressure.

Treatment

A $\frac{1}{8}$ to $\frac{1}{4}''$ *lateral heel wedge* can help.

Oral *antiinflammatories* may be used.

Steroid injection can be offered; success rate is good. The procedure is analogous to that for posterior tibialis tendonitis, discussed above.

Anterior Tibialis and Extensor Tendonitis (*Uncommon*)

Diagnosis

Tenderness occurs along these tendons in the anterior ankle and dorsum of the foot (Figure 17-2).

Be sure you are not dealing with a rare infectious tenosynovitis (distinguished by redness, warmth and acute onset).

Pathophysiology

Most commonly this is a pressure phenomenon, resulting from wearing shoes that are too small or tied too tightly.

Sometimes strain is excessive because too-high heels are worn, leading to too much contraction of these muscles (to prevent a slapping gait) and to stretch of the tendons.

Or it can sometimes be posttraumatic (following foot or ankle sprain).

The extensor hallucis longus tendon popping over the anterior talus is the most common cause of the complaint of "snapping ankle." The diagnosis can be confirmed by palpating there as the ankle is actively circumducted.

Treatment

Local *pressure* must be *alleviated* or the heel *lowered,* depending on the suspected mechanism.

Oral *antiinflammatory medications* can be used.

Steroid injection may be offered; success rate is good. The procedure is analogous to that for posterior tibialis tendonitis, discussed above.

Degenerative Arthritis (Osteoarthritis) of the Subtalar Joint (*Common*)

Diagnosis

An older patient or one with significant foot pronation or previous trauma complains of chronic pain which is worse after much weight-bearing.

Tenderness is found over the sinus tarsi and the subtalar joint line (Figure 17-2).

Often passive range of motion of the joint is painful, and there sometimes is crepitus.

Changes are often seen on X ray, though it may appear normal early in the course.

Pathophysiology

The process of chronic foot strain described in Chapter 18 alters the mechanics of this joint between the talus and calcaneus and leads to accelerated degeneration.

See also Chapter 22.

Treatment

The measures recommended in Chapter 18 for patients with *chronic foot strain* must be employed.

Oral *antiinflammatory* medication may be of benefit.

Injection of *steroid* into the joint may be offered. See Figure 17-5.

Degenerative Arthritis (Osteoarthritis) of the Ankle (*Common*)

Diagnosis

An older patient or one with previous trauma will complain of chronic pain which is worse after weight-bearing.

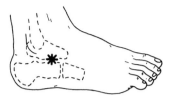

LATERAL

Fig. 17-5. How to aspirate or inject a subtalar joint. Passively invert the ankle, and palpate about a centimeter in front of and just inferior to the tip of the lateral malleolus. The depression felt there is the *sinus tarsi*, through which the subtalar joint can most easily be entered. The rest of the technique is exactly as described in Figure 17-3 (including the cautions *re* infection and steroid instillation).

There may be variable diffuse tenderness, and variable pain with range of motion.

Slight joint swelling can occur in an acute exacerbation, but if the joint is red or warm it must be tapped to rule out infection.

The X ray may be normal early but is characteristic later.

Pathophysiology

See Chapter 22.

Treatment

See Chapter 22.

If the patient is obese, *weight loss* is helpful. Also, *good shoes* as described in Patient Handout 18-1 should be worn.

Technique of *injection* is shown in Figure 17-3.

ANKLE SPRAIN

Examination

Look for *tenderness*. It is usually limited to the anterior talofibular ligament in first-degree sprains, but often extends to the fibulocalcaneal ligament and even the posterior talofibular ligament in more severe sprains (Figure 17-6).

Isolated tenderness over *only* the posterior aspect of the lateral ankle may be indicative of a traumatic subluxation of the peroneal tendons rather than a ligamentous sprain. The tendons have usually spontaneously relocated by the time the patient is seen. The patient is best referred to an orthopedist for management, which may consist of padding, casting or surgery.

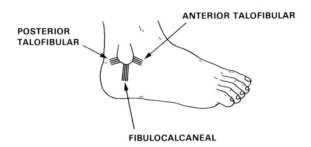

Fig. 17-6. The lateral ligaments of the ankle.

If there is tenderness and/or swelling over the *medial aspect* of the ankle (deltoid ligament) or the anterior aspect of the ankle, *severe injury* can be assumed and the patient is best referred to an orthopedist.

Soft tissue swelling is localized in first-degree sprains and is more generalized in higher-degree injuries.

The presence of *ecchymosis* signifies at least some ligamentous tearing, and thus a second-degree injury at the minimum.

Test for *stability* by performing the anterior drawer maneuver (Figure 17-7); good muscular relaxation is necessary for accuracy, and results must always be compared with the uninjured limb because there is great variation from individual to individual. Significant asymmetric laxity implies a third-degree injury.

See if lateral *compression* of the *distal tibia* and *fibula* is painful. If so, the patient has an injury to the distal interosseous ligament and should be referred to an orthopedist for treatment. Test distal *strength, sensation,* and *capillary filling.*

X Ray

One should be obtained in all but the most minor injuries.

It will of course be normal in a sprain of the lateral ligaments. (A small avulsion fracture implies total ligamentous disruption and therefore should be treated as a third-degree sprain. Any other fracture is considered beyond the scope of this text.)

If on the AP film a clear space is visible at all points between the tibia and fibula, the patient may well have a disruption of the distal interosseous ligament. If a view of the uninjured leg confirms asymmetry, the patient should be seen by an orthopedist (Figure 17-8).

Since even third-degree sprains are now generally treated conservatively (except in athletes) and since asymmetry in anterior distraction of the ankle is a more

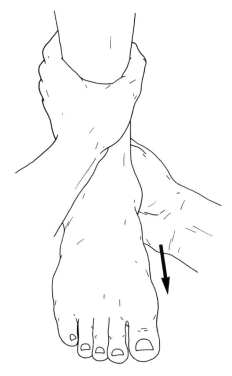

Fig. 17-7. The anterior drawer test (in the ankle). Good muscle relaxation is essential. Always compare to the uninjured side.

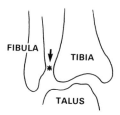

Fig. 17-8. Interosseous ligament disruption. Clear space at all points between the tibia and fibula on the A-P X ray indicates a possible disruption of the interosseous ligament.

reliable sign of ligamentous disruption than excessive inversion, the taking of stress inversion X rays is usually unnecessary.

First Degree Sprain (*Very Common*)

Diagnosis

The patient gives a history of a misstep and inversion stress.

On examination, tenderness and swelling are limited (usually to the area of the anterior talofibular ligament), there is no ecchymosis, and the anterior drawer test is negative.

The X ray is negative.

Pathophysiology

This is a stretching of a ligament, usually the anterior talofibular, by forced inversion with the foot in mild plantar flexion.

Treatment

Acutely, advise *elevation, compression* and *ice*.

Weight-bearing should be *limited* with progression over a few days as tolerated. See Patient Handout 17-1.

Second Degree Sprain (*Common*)

Diagnosis

The patient may report that he felt a tearing sensation as he made his misstep.

On examination, fairly diffuse swelling is found, with tenderness over the lateral ligaments and usually some ecchymosis. The anterior drawer test may show a bit of asymmetry with the uninjured side, but not much.

The X ray is negative.

Pathophysiology

Some of the ligamentous fibers are torn, but there is not a complete disruption. The force is usually great enough to injure the adjacent ligaments to some degree as well.

Treatment

Ice packs, *compression, immobilization,* and *elevation* should be used for the first 48 to 72 hours.

The patient should be issued *crutches* (and taught how to use them), and no weight bearing should be allowed. (A short leg walking cast may be used if the patient is elderly or is unable or unwilling to follow the outline shown here; see the discussion of treatment under "third-degree sprain" below.)

A new device that is proving very useful in athletes and other reliable patients is the *pneumatic ankle brace,* in which rigid parts prevent inversion/eversion motion while allowing plantar flexion/dorsiflexion, and the form-fitting pneumatic inner helps pump out edema.

He or she should begin plantar flexion/dorsiflexion *exercises* after the first 48 hours or so. This spares the damaged areas, but maintains joint range of motion and proprioceptive sensation. (The latter may be impaired permanently after a mere ten days in a cast.) It also helps to prevent atrophy of at least some of the muscles.

The patient should return for a recheck after 5 to 10 days, and if progress is adequate (judged by a significant decrease in swelling and pain) should begin weight-bearing again. An elastic bandage or anklet is helpful so that the patient keeps the injury in mind, but should not be thought of as providing any significant protection from reinjury.

At two to three weeks postinjury, *circumduction exercises* should be started, and then continued for several weeks at least.

The patient should be told to avoid walking or running on sloped or uneven surfaces for several weeks, to elevate and ice the ankle as needed, and to tape the ankle before athletic endeavor.

(This treatment program is outlined in Patient Handout 17-2).

Third Degree Sprain (*Uncommon*)

Diagnosis

The patient will usually report a painful tearing sensation at the moment of injury.

There may be less subjective pain than with a second-degree sprain, because there is no injured structure under tension.

On examination, there is diffuse swelling and tenderness of the lateral ankle, as well as ecchymosis.

The anterior drawer test reveals laxity.
The X ray is normal or shows a small avulsion chip.

Pathophysiology

This is a complete disruption of one or more of the lateral ligaments (usually the anterior talofibular), with associated lesser degrees of injury to surrounding structures.

Treatment

Ice, elevation, and *compression* in a *boot* or *Jones dressing* should be used for the first 48 to 72 hours and no weight bearing should be allowed. *Crutches* should be issued.

When the swelling has abated sufficiently, a *short leg walking cast* should be applied for four to six weeks. A *pneumatic ankle brace* is an available alternative for reliable patients.

When the cast is removed the patient should be issued crutches and immediately begun on circumduction exercises (as shown in Patient Handouts 17-1 and 2). The crutches may be discontinued after about a week out of plaster, but athletic activity should be prohibited for at least another month.

The patient should be told to avoid walking or running on uneven surfaces and to use elevation and ice as needed, and taught how to tape the ankle (Patient Handout 17-3).

Persistent pain is usually due to a *posttraumatic tendonitis* (see ANKLE PAIN), but if this is not the case, suspect an unrecognized *talar dome fracture* and refer the patient to an orthopedist.

Some orthopedists still feel that *primary surgical repair* of third-degree sprains is advisable; this should be considered especially if the patient is an athlete.

ANKLE INSTABILITY

Evaluation

This complaint is usually due to lax ligaments (often with weak musculature) following a sprain, but can be due to a loose body within the joint. If the latter is suspected or noted on X ray, referral to an orthopedist is recommended.

Laxity of the Lateral Ligaments (*Common*)

Diagnosis

The patient gives a history of recurrent inversion sprains.
Examination reveals some laxity with inversion stress and/or the anterior drawer maneuver (Figure 17-7).

Pathophysiology

Congenitally lax ligaments may be a contributing factor, but previous injury is usually more important.

Treatment

The patient should as much as possible *avoid* walking or running on *uneven surfaces* and activities which require a lot of *jumping* or *sudden starts, stops,* and *direction changes.*
Circumduction exercises must be performed faithfully.
The patient should be taught how to *tape* the ankle prior to participating in athletics; or a *pneumatic ankle brace* can be prescribed.
A one-fourth inch lateral *heel wedge* may be employed.
Referral for *surgery* is sometimes necessary despite all these measures.

ANKLE LUMPS

GANGLIA are common about the ankle; see the discussion in Chapter 8, which is equally applicable here.

Patient Handout 17-1

Caring for Your Sprained Ankle (First-Degree)

Your doctor has determined that your ankle sprain is *first-degree*. That's good: it means the ligaments are only stretched, and none of them are torn. These injuries usually heal quite well.

You should put an *ice pack* on the area for twenty minutes every few hours for the first couple of days. (Never put ice directly on the skin.) Keep your ankle elevated. An elastic bandage may help a bit to keep the swelling down. Limit walking on it; as the days go by and the pain and swelling decrease, you can slowly resume your normal activity.

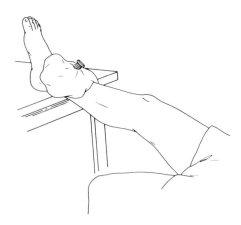

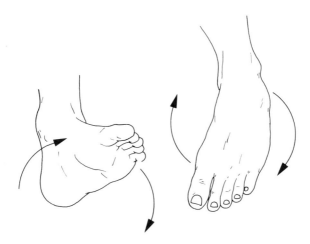

Stay off high heels or sloped or uneven surfaces for a few weeks. Another good idea is to do circumduction exercises, where you move your ankle in a circle ten times in one direction and then ten times in the other, several times a day. This will keep the muscles around your ankle strong. Don't start them until all the pain and swelling are gone.

If you take part in athletics you may also want to learn how to *tape* your ankle for extra protection: ask your doctor for instructions.

Patient Handout 17-2

Caring for Your Sprained Ankle (Second-Degree)

Your doctor has determined that your ankle sprain is *second-degree*, which means that some fibers of the ligaments on the outside of your ankle are torn, but not all of them. This injury often heals quite well if you follow advice, but can lead to prolonged pain and a weak ankle if you take it lightly.

For the first two or three days after your sprain, put an *ice pack* on the area for twenty minutes every hour or two. (Never put ice directly on the skin.) Keep your leg elevated on a table or hassock when sitting, and on a pillow or two when lying down. Use whatever dressing your doctor has given you, or an elastic wrap if he or she says it's O.K.

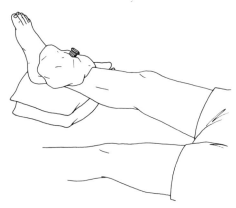

You will have been given crutches. If you don't know how to use them properly, be sure to ask. These have been given to you for a reason: you must put no weight on your injured ankle until your doctor says it's all right. In this way the tissues have a chance to heal properly. So use the crutches to get around.

After two or three days, you should begin doing the exercise shown. Move your ankle up and down only, *not* side to side at all. Do it a dozen times an hour. This is important so that you don't lose feeling in the ankle joint, and it will help get rid of some of the swelling, too.

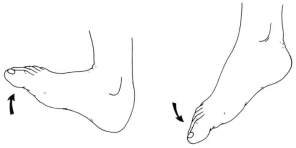

When your doctor says you can go back to putting weight on your foot, do it gradually, using your crutches less each day for a few days. Certainly stay off high heels and sloped or uneven surfaces, and don't do any running yet. About this time, you should begin to do circumduction exercises, where you move your ankle in a circle ten times one way and then ten times the other way each hour. Keep these exercises up for four to six weeks at least. Your doctor will advise you when you can go back to athletic activity. When you do, it would be wise to *tape* your ankle: your doctor will advise you how.

If you have *persistent* pain or swelling, it doesn't necessarily mean anything bad, but be sure to let your doctor know.

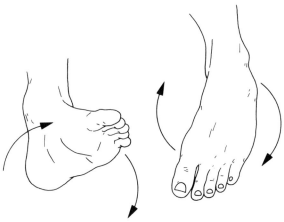

Patient Handout 17-3

How to Tape Your Ankle

Use heavy cloth tape, $\frac{1}{2}$-in. wide.

Start on the inside of your leg, above your ankle, and go down under your heel and up the outside of your leg to above your ankle. The trick is to keep your ankle turned outward and maintain tension on the tape throughout. Add more strips, each one about half overlapping the one in front. Then anchor these strips with some overlapping strips going ¾ of the way around the leg.

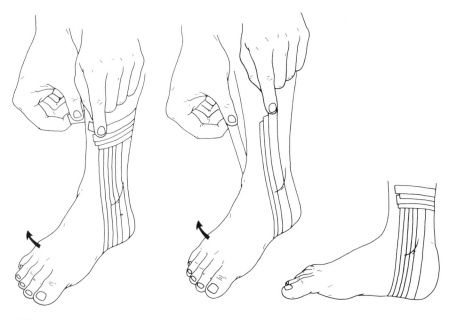

18
The Foot

Foot and heel problems in *infants* and *children* are covered in Chapter 19. Foot problems in *runners* are specifically discussed in Chapter 20. *Numbness and weakness* of the foot are covered in Chapter 13. *Foot cramps* are covered in Chapter 21. Problems around the *ankle* are the subject of Chapter 17. *Contracture* of the plantar fascia is analogous to Dupuytren's contracture in the hand. See Chapter 9.

For the anatomy of the foot see Figure 18-1.

HISTORY

Ask about the *duration* of symptoms.

Has there been any *specific trauma, strain,* or *overuse?*

What types of *shoes* are worn? Has there been a recent change?

Have the patient show you the *location* of the pain.

Is there any *paresthesia* or *weakness?*

Inquire about *precipitating* and *relieving activities* and types of *footwear.*

Is there any previous history of gout or other joint problems?

Is the patient *diabetic?*

EXAMINATION

Look for specific sites of *tenderness,* and if any are found refer to the appropriate section in this chapter or the chapter on the ankle.

Test *range of motion* of the various joints.

Observe the patient while he or she is standing or walking to detect *loss of longitudinal arch* and *excessive pronation.*

Look at the patient's *shoe,* especially noting localized wear, poor fit, or out-of-the-ordinary shape.

Check *neurovascular status.*

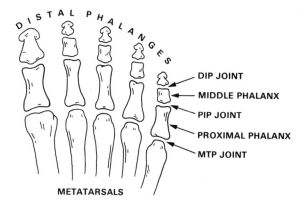

DIP JOINT

MIDDLE PHALANX

PIP JOINT

PROXIMAL PHALANX

MTP JOINT

METATARSALS

BONES AND JOINTS OF FOREFOOT

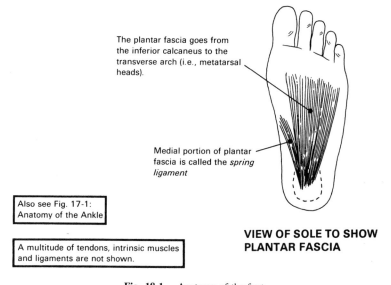

The plantar fascia goes from the inferior calcaneus to the transverse arch (i.e., metatarsal heads).

Medial portion of plantar fascia is called the *spring ligament*

Also see Fig. 17-1: Anatomy of the Ankle

A multitude of tendons, intrinsic muscles and ligaments are not shown.

VIEW OF SOLE TO SHOW PLANTAR FASCIA

Fig. 18-1.   Anatomy of the foot.

DIFFUSE FOOT PAIN

Evaluation

See the recommended history and examination in the beginning of this chapter. Note that poorly localized foot pain with or without neurologic symptoms may be due to any of the conditions covered in Chapter 13. Specifically, diffuse pain in the heel or sole can be secondary to the TARSAL TUNNEL SYNDROME.

Acute Foot Strain (*Common*)

Diagnosis

Diffuse foot pain follows unaccustomed standing or walking.
Examination is normal except possibly for mild tenderness in various locations.

Treatment

A period of *rest* and *elevation* is usually all that is needed.
Ice packs or *antiinflammatories* may be helpful in more serious cases.

Chronic Foot Strain (Hyperpronating foot, flat foot, pes
planus) (*Very Common*)

Diagnosis

The patient complains of poorly localized aching in the foot, worse with
prolonged standing or walking.
Or it may be the underlying cause of several of the more specific complaints
discussed elsewhere in this chapter.
See Figure 18-2.

Pathophysiology

Congenital flat feet can predispose to chronic foot strain later in life.
More commonly, however, the process is due to shoes that allow loss of tone
of the supinator and intrinsic flexor muscles and tend to raise the outside of the
foot.
The talus begins to slip downward and distally. This in turn causes eversion of
the calcaneus and lowers the longitudinal arch, which in turn leads to pronation
of the forefoot and therefore to broadening and lowering of the transverse
metatarsal arch.
With these changes, the weight transfer process during walking shifts from
*heel → lateral border of foot → metatarsal heads → push-off by ball of foot and
big toe* to one of a *straight line from heel to the middle metatarsal heads*.

Treatment

Shoes should be worn that have a snug counter, good longitudinal arch support,
and a well-fitting last. The shoe should also have ample room for the toes and a
low heel. See Patient Handout 18-1.

An additional longitudinal *arch support* and *medial heel wedge* should be added when needed. (The latter is contraindicated if the patient is predisposed to recurrent inversion sprains of the ankle.)

Patients who are young and motivated should practice *toe gripping* when walking until it becomes automatic, and can be taught *exercises* to strengthen the supinator and intrinsic flexor muscles. See Patient Handout 18-2.

Rest, analgesics, and *warm foot baths* should be used as symptoms warrant. If the patient is obese, *weight loss* can be of major benefit.

Specific treatment for the various specific *sequelae* of the chronically strained foot are discussed under the specific syndrome headings.

Pes Cavus (Clawfoot, High Arch) (*Uncommon*)

Diagnosis

By inspection (see Figure 18-2).

Pathophysiology

These feet are rigid and absorb shock poorly. They can be asymptomatic or may lead to diffuse foot pain or to metatarsalgia and clawing of the toes.

The condition may rarely be associated with underlying neurologic disease.

Treatment

Be sure that the patient wears flexible shoes so as not to make the foot's shock absorption capacity any worse.

In mild cases, a *small heel lift* and the use of a fitted *arch support* are helpful.

Daily repetitive *passive stretching* of the proximal extensors and distal flexors of the toes may help prevent deformity there.

Tarsal Coalition (*Uncommon*)

Diagnosis

The patient usually complains of diffuse foot pain.

Examination will reveal restricted passive motion between the heel and the ankle. The arch will flatten with weight-bearing.

A bony bridge will usually be visible on X ray between the calcaneus and either the talus or the navicular. (Rarely, the bridge is cartilagenous and not visible on X ray.)

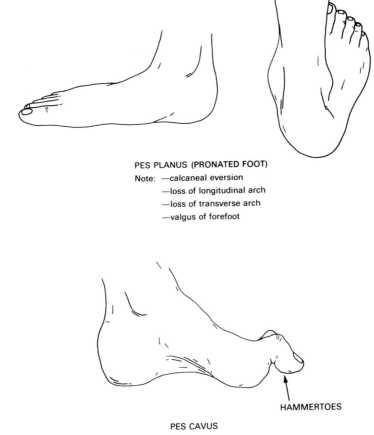

PES PLANUS (PRONATED FOOT)
Note: —calcaneal eversion
       —loss of longitudinal arch
       —loss of transverse arch
       —valgus of forefoot

HAMMERTOES

PES CAVUS

Fig. 18-2.   Pes planus and pes cavus.

Pathophysiology

The condition is congenital.

The rigidity caused by the coalition leads to poor shock absorption and thus diffuse pain and degenerative change. The restricted motion usually results in spasm of the peroneal muscles.

Treatment

Shoes must be as *flexible* as possible.

If this, and prn rest and antiinflaminatory medication, do not give sufficient relief, referral for *surgery* is warranted.

Diffuse Degenerative Arthritis (Osteoarthritis) of the Foot (*Uncommon*)

Diagnosis

The patient complains of diffuse pain that is worse after walking or standing. Often degenerative changes are more advanced in the subtalar and first MTP joints (discussed separately below and in Chapter 17). Changes are usually visible on X ray.

Pathophysiology

This is usually a sequel of chronic foot strain, but may also result from the poor shock absorption of a rigid foot.
See also the discussion of degenerative arthritis in Chapter 22.

Treatment

The *measures discussed under CHRONIC FOOT STRAIN,* or the appropriate other underlying entity, should be employed.
Use rest, heat, and *oral antiinflammatory medications* as needed.
If a specific joint is contributing greatly to the symptoms, *steroid injection* may be offered.

HEEL PAIN

Evaluation

See Figure 18-3.
If no tenderness is found the pain may be referred from DEGENERATIVE ARTHRITIS OF THE SUBTALAR JOINT, covered in Chapter 17.
Acute inflammation in the heel area can be a manifestation of REITER'S SYNDROME (Chapter 23), or rarely, of GOUT (Chapter 24).
For problems around the ankle, see Chapter 17.

Plantar Fasciitis (''Heel Spur'') (*Common*)

Diagnosis

The patient complains of anterior heel pain often with radiation forward into the sole or arch and worse with prolonged weight-bearing.

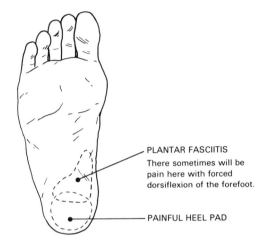

PLANTAR FASCIITIS
There sometimes will be
pain here with forced
dorsiflexion of the forefoot.

PAINFUL HEEL PAD

Fig. 18-3. Sites of tenderness: inferior heel. (Tenderness may be anywhere within the areas shown.)

Commonly there will be a good deal of pain for the first few steps after a period of non-weight-bearing (upon arising in the morning, for example).

On examination, there is tenderness anywhere along the plantar fascia insertion (Figure 18-3).

Often pain can be increased by forcibly dorsiflexing the toes and forefoot, thus stretching the plantar fascia.

X ray may or may not show a spur. The spur, if present, is evidence of long-term excessive strain on the insertion of the plantar fascia, but it is not in itself the etiology of the patient's pain.

Pathophysiology

Inflammation occurs where the plantar fascia inserts into the calcaneus due to excessive stretch from chronic foot strain (excessive pronation and a lowered longitudinal arch) or in a cavus foot.

Treatment

The *measures discussed in CHRONIC FOOT STRAIN* must be applied, especially the use of a longitudinal arch support.

Be sure the sole of the shoe isn't too stiff.

A course of oral *antiinflammatory medication* can be helpful.

Steroid injection may be offered. Success rate is only fair. See Figure 18-4.

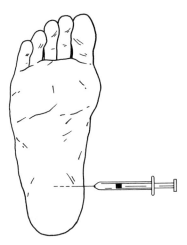

Fig. 18-4. Injection for plantar fasciitis. After prepping the area and spraying with ethyl chloride, inject toward the point of maximal tenderness from the medial approach. (This is much less painful than the inferior approach.) Use a long 25-gauge needle and inject about a cc of steroid with a cc of local anesthetic. See pp. 23–27 for further details and precautions.

Painful Heel Pad ("Stone Bruise") (*Uncommon*)

Diagnosis

Pain and tenderness occur diffusely under the entire heel or the posterior part of it.

Pathophysiology

The heel pad loses its elasticity as the body ages, and acute or repetitive strain or trauma can cause the entire layer of fatty and fibrous elastic tissue to become inflamed.

Treatment

The heel should be *raised* a bit to transfer more weight anteriorly. Note that this is the opposite of what is needed for plantar fasciitis.
A *sponge rubber pad* can be helpful.
Oral *antiinflammatory medications* or *injection of steroid* may be tried.

PAIN IN THE ACHILLES TENDON AREA

Evaluation

See Figure 18-5.

Posterior Calcaneal Bursitis (Pump Bump) (*Uncommon*)

Diagnosis

A tender bump is found over the achilles tendon (Figure 18-5).

Pathophysiology

A bursa forms and becomes irritated from the upper border of a heel counter that is too firm and/or convex.

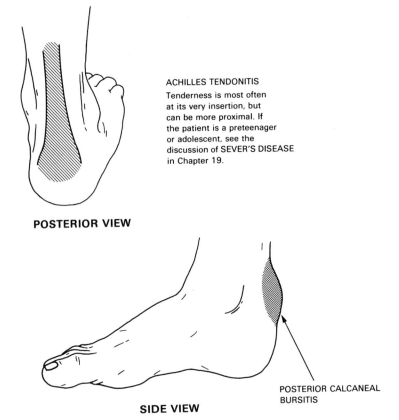

ACHILLES TENDONITIS
Tenderness is most often
at its very insertion, but
can be more proximal. If
the patient is a preteenager
or adolescent, see the
discussion of SEVER'S DISEASE
in Chapter 19.

POSTERIOR VIEW

SIDE VIEW

POSTERIOR CALCANEAL
BURSITIS

Fig. 18-5.   Sites of tenderness: posterior heel.

Treatment

The *offending shoe* must be *avoided*, with or without the temporary use of shoes with a completely open heel to allow some resolution.

If severe, a small amount of *steroid* may be *injected* into the bursa. Never inject into the achilles tendon itself. See Chapter 3 for further details and precautions.

In chronic cases, referral for *surgical excision* can be offered.

Achilles Tendonitis (*Common*)

Diagnosis

Pain and tenderness are located at the achilles tendon insertion (Figure 18-5) or proximal to it.

Pathophysiology

Excessive stretch occurs there from unaccustomed activity, especially with much uphill walking, or from a new shoe with a heel lower than the patient had been wearing.

Treatment

In long term problems or in athletes, *stretching exercises* are most important. See Chapter 20.

A small *heel lift* should be placed in both shoes (about one-quarter inch or so).

Rest, hot or *cold compresses,* and oral *antiinflammatory medications* can be used in acute cases.

Injection is *contraindicated* because the subsequent weakening of the tendon can predispose to rupture, which is catastrophic.

PAIN IN AND AROUND THE BALL OF THE FOOT

Evaluation

Sudden atraumatic onset of pain and swelling in the first MTP joint is almost pathognomonic of GOUT. See Chapter 24.

Look for a HALLUX VALGUS condition (Figure 18-6) and a tender or nontender thickening on the medial aspect of the joint. This is a BUNION.

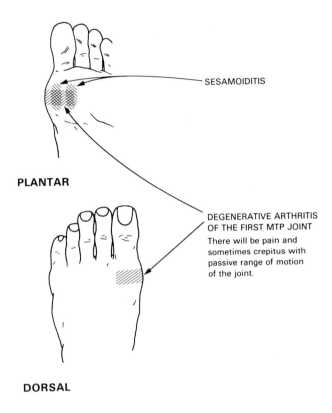

SESAMOIDITIS

PLANTAR

DEGENERATIVE ARTHRITIS
OF THE FIRST MTP JOINT
There will be pain and
sometimes crepitus with
passive range of motion
of the joint.

DORSAL

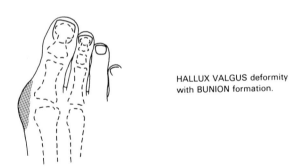

HALLUX VALGUS deformity
with BUNION formation.

Fig. 18-6.   Sites of tenderness: ball of foot.

Hallux Valgus and Bunion (*Common*)

Diagnosis

HALLUX VALGUS refers to lateral angulation of the great toe at the MTP joint.

A BUNION is a chronically thickened and continually or intermittently inflamed bursa on the medial aspect of the joint.

The analogous condition occurring at the fifth MTP joint is known as a Taylor's bunion or bunionette.

Pathophysiology

Although some congenital malformations (metatarsus primus varus or a deformed first metatarsal head) predispose to hallux valgus, the usual direct cause is the wearing of shoes with narrow toe areas.

Excessive pressure is then placed on the medial aspect of the first MTP joint, with the resultant development of a thickened, inflamed bursa.

Treatment

The wearing of *shoes with wide toeboxes* should be advised, and a *donut pad* (Figure 18-8) or lamb's wool may be helpful.

Splints and exercises to halt or reverse the development of hallux valgus are usually not helpful because the bunion does not become painful, and thus the patient does not present, until the deformity is well-established.

Referral to a podiatrist or orthopedist for consideration of *surgery* is often necessary.

Degenerative Arthritis (Osteoarthritis) of the First MTP Joint (*Common*)

Diagnosis

Variable diffuse joint line tenderness is found, as well as variable pain with range of motion of the joint, often with crepitus (Figure 18-6).

The X ray may be normal early.

A uric acid level should be drawn to help exclude the diagnosis of gout.

Pathophysiology

Accelerated wear and tear to this joint occurs from excessive flexion/extension.

It is common in people who wear high heels, in runners, in soccer players, and in dancers.

Treatment

Stiffening the sole (with a stiff insert or with steel-shank shoes) helps by decreasing the motion at the joint.

It is important that *low heels* be worn.

Oral *antiinflammatory medications* or intraarticular *steroid injection* (Figure 18-7) may help the patient over an acute exacerbation.

Eventually the joint may spontaneously fuse, a condition known as *hallux rigidus*.

Referral for *surgery* may be necessary if the measures noted above do not result in sufficient relief.

Sesamoiditis (*Rare*)

Diagnosis

Pain and tenderness are limited to the area under the sesamoid bones (Figure 18-6).

Pain often results if the involved sesamoid is compressed upward while the great toe is passively dorsiflexed.

These bones are often bipartite, and this should not be confused on X ray with a fracture.

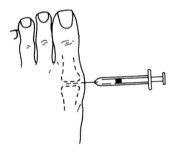

Fig. 18-7. Injection into the first MTP joint. Feel the joint line over the dorsal–medial aspect with your fingertip. Moving the base of the big toe up and down can help localize it. After sterile prep and ethyl chloride spray, enter the joint with a short 25-gauge needle and inject a cc of steroid and a cc of local anesthetic without epinephrine. See pp. 23–27 for further details and precautions.

Pathophysiology

Pain is due to excessive pressure or to roughening of the superior surface of the sesamoid, with consequent friction as the toe is flexed and extended.

Treatment

A *stiff sole, low heel,* and *donut pad* (Figure 18-8) are helpful.
Steroid injection with a #25 needle from the medial side may be offered. See Chapter 3 for details and precautions.

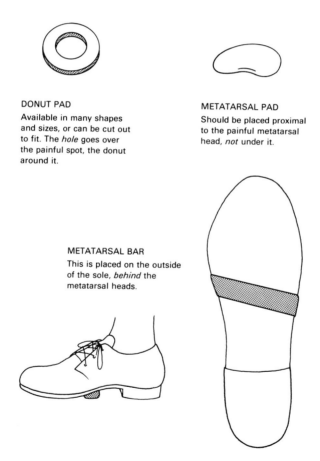

DONUT PAD
Available in many shapes and sizes, or can be cut out to fit. The *hole* goes over the painful spot, the donut around it.

METATARSAL PAD
Should be placed proximal to the painful metatarsal head, *not* under it.

METATARSAL BAR
This is placed on the outside of the sole, *behind* the metatarsal heads.

Fig. 18-8. Therapeutic devices for the foot.

FOREFOOT PAIN

Evaluation

See Figure 18-9.

Metatarsalgia (*Common*)

Diagnosis

Tenderness is located under the second (sometimes the third) metatarsal head, often with callus formation as well.

Pathophysiology

Chronic foot strain changes weight-bearing in such a way as to place more pressure at this point. A cavus foot can do the same.

Often a change in shoe (especially to a higher heel) will be the precipitant.

Treatment

A *low heel* should be worn.

The *measures described for CHRONIC FOOT STRAIN* are helpful.

A *metatarsal pad* (Figure 18-8) is useful.

In more chronic or refractory cases a *metatarsal bar* (Figure 18-8) may have to be tried.

In severe cases a course of an *antiinflammatory medication* or local *steroid injection* should be offered.

Referral to a podiatrist for more precise biomechanical correction should be considered in refractory cases, with surgery as a last resort.

Morton's Neuroma* (Interdigital Neuroma) (*Common*)

Diagnosis

Tenderness is *between* the third and fourth (or sometimes the second and third, or rarely other) metatarsal heads (Figure 18-9).

---

*Not to be confused with MORTON'S FOOT, which is a congenitally short first metatarsal and is named after a different Morton.

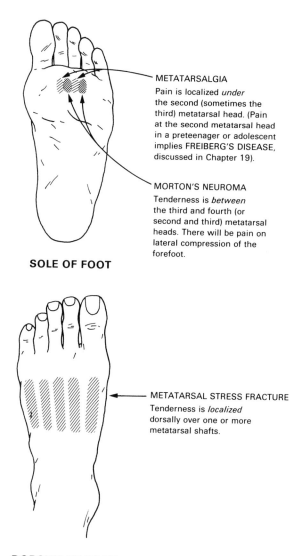

SOLE OF FOOT

METATARSALGIA
Pain is localized *under*
the second (sometimes the
third) metatarsal head. (Pain
at the second metatarsal head
in a preteenager or adolescent
implies FREIBERG'S DISEASE,
discussed in Chapter 19).

MORTON'S NEUROMA
Tenderness is *between*
the third and fourth (or
second and third) metatarsal
heads. There will be pain on
lateral compression of the
forefoot.

METATARSAL STRESS FRACTURE
Tenderness is *localized*
dorsally over one or more
metatarsal shafts.

DORSUM OF FOOT

Fig. 18-9. Sites of tenderness: forefoot.

Pain is produced by lateral compression of the forefoot.
Sometimes there is numbness in the interspace distal to the neuroma.
The patient will report that relief is obtained more by removing the shoe than
by getting off the foot.

Pathophysiology

This is a fusiform swelling of the interdigital nerve, usually secondary to
compression by the metatarsal heads in a chronically strained foot or an overly
tight shoe.

Treatment

Shoes must be *wide* enough so that further irritation of the nerve is avoided.
The other *measures discussed under CHRONIC FOOT STRAIN* may be tried.
Steroid injection may be offered. Success rate is fair. See Figure 18-10.
Often referral to a podiatrist or orthopedist for *excision* becomes necessary.

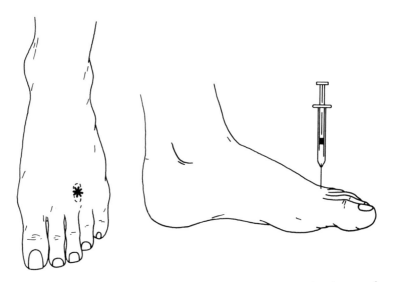

Fig. 18-10. Injection for Morton's neuroma. Localize the neuroma on the *plantar* surface between the metatarsal heads; now place your finger on the *dorsal* surface directly superior to the
neuroma. After prep and ethyl chloride spray, inject *directly downward* with a 25-gauge needle about
a cc of steroid and a cc of local anesthetic without epinephrine. See pp. 23–27 for further details and
precautions.

Stress Fracture (March Fracture) of the Metatarsal Bone
(*Uncommon*)

Diagnosis

Forefoot pain comes on after overuse (e.g., a runner who has increased his mileage or a new soldier forced to take long marches) and decreases somewhat with rest.

On examination, well localized tenderness is found on one or more of the metatarsal shafts on the dorsum of the foot (Figure 18-9).

Pathophysiology

See p. 4.

Treatment

Decrease weight-bearing activity. If stress on the injured bone continues unabated, the lesion may extend and become a real through and through fracture.

A short leg walking *cast* for two to four weeks is unnecessary but may be used for symptomatic relief.

Analgesia should be used as needed.

TOE PROBLEMS

Hammertoes (*Common*)

Diagnosis

By inspection (Figure 18-2).

Pathophysiology

May be secondary to pes cavus or may be due to the long term wear of shoes and/or socks that are too short and crowd the toes into flexion.

The weakening of the intrinsic flexors that comes with shoe wear may be a contributing factor in those patients whose anatomy predisposes them to the problem.

Treatment

Shoes with a *high toe box* should be obtained.
Treat the complications, metatarsalgia and corns, as discussed under those headings.
Referral for *surgery* may become necessary.

Ingrown Toenail (*Very Common*)

Diagnosis

Acute or chronic inflammation occurs at the nail margin.

Pathophysiology

If the corner of the nail is trimmed too aggressively, the nail can grow into the overhanging skin.
A very convex nail predisposes to the problem.

Treatment

In acutely inflamed cases, treat with an *antibiotic* effective against staphylococcus as well as with *elevation, open shoes,* and twice daily *warm soaks.*

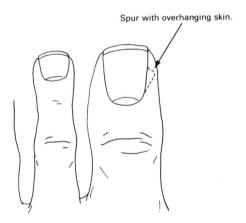

Spur with overhanging skin.

Fig. 18-11. Removing a nail spur. After obtaining informed consent and anesthetizing the toe analogously to Figure 9-9, bluntly dissect the spur out from under the overhanging soft tissue and cut it off using scissors.

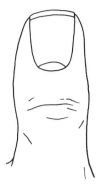

Fig. 18-12. Proper toenail trimming. Note how nail is cut straight across and how corners protrude just a bit instead of digging into the skin.

If the edge of the nail can be pried out of the skin, *daily insertion of a piece of cotton* between the nail and the skin should be performed by the patient until the corner of the nail has grown past the overhanging tissue. (This should be done after soaking, and with a blunt instrument such as the edge of a tongue blade.)

If there is a *spur* protruding into the soft tissues (often from previous efforts at partial nail removal), the method shown in Figure 18-11 may be tried to *remove* it.

If it is decided that a strip of nail must be removed, the corresponding area of matrix must be obliterated or the problem will quickly recur. A variety of methods are used; consult a text on outpatient surgery, or refer the patient.

To prevent recurrence, patients should be instructed to wear wide enough shoes and always trim their nails straight across (Figure 18-12).

KERATOSES

Evaluation

See Figure 18-13.

Callus (*Very Common*)

Diagnosis/Pathophysiology

This is a diffuse or circumscribed thickening of the skin resulting from excessive pressure or friction.

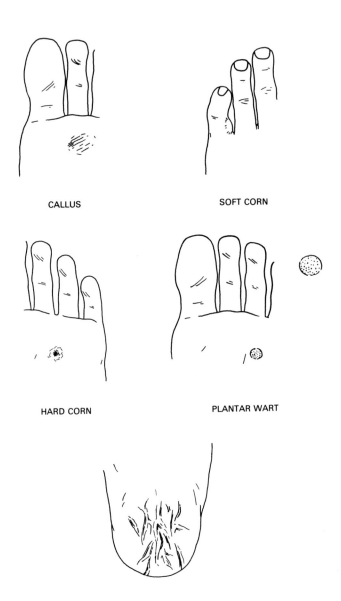

CALLUS   SOFT CORN

HARD CORN   PLANTAR WART

KERATODERMA PLANTARIS

Fig. 18-13.   Keratoses.

Treatment

The *underlying cause* of the abnormal pressure should be dealt with.
Shoes should not cause friction at, or put unnecessary pressure on, the callused area.
A *donut pad* (Figure 18-8) should be used.
The callus may be *scraped* (*never* cut into) with a blade or *rubbed* with a pumice stone after *soaking* to soften it.

Soft Corn (*Common*)

Diagnosis/Pathophysiology

These are found between the toes (usually the fourth and fifth). They are caused by pressure from an adjacent toe.

Treatment

As for a callus (above).
Cotton or *lamb's wool* placed between the toes is often helpful.
Forty percent *salicylic acid plasters* may be applied for 24 hours every few days, always after soaking.
Referral for *surgical excision* of underlying bony prominences may become necessary.

Hard Corn (Neurovascular Corn) (*Common*)

Diagnosis/Pathophysiology

It is sharply demarcated and appears over a bony prominence.

Treatment

As for a callus or soft corn.

Plantar Wart (*Common*)

Diagnosis

It is sharply demarcated, and not necessarily found over a bony prominence.
Scraping will reveal black dots that are actually capillary endings.

Skin lines go around a wart; they go through a callus or corn.
Warts are painful with lateral compression, while calluses and corns generally are not.

Pathophysiology

Caused by the same papilloma virus that causes warts elsewhere; because of pressure they grow inward rather than outward.

Treatment

Surgery and overly aggressive treatment that will leave a scar should be avoided. (A scar will be permanently painful; at least the wart will eventually go away if the patient holds out long enough.)
Soaking followed by the application of a keratolytic such as 40% *salicylic acid plaster* sometimes works.
Liquid nitrogen or *paring followed by the application of trichloroacetic acid* to the base are other methods available.

Keratoderma Plantaris (*Uncommon*)

Diagnosis

Diffuse thickening of the skin under the heel is found, with the development of often-painful fissures.

Treatment

Advise the patient to use *warm soaks* followed by application of a *keratolytic*.
Antibiotics may have to be used if the fissures become infected.

FOOT LUMPS

GANGLIA are common in the ankle and foot; see the discussion in Chapter 8, which is equally applicable here. Many different small ACCESSORY BONES have been described, and are usually not significant except for the pressure effects they may cause. Diagnosis is by X ray and treatment by neglect, padding or surgery as necessary. (But see the discussion of the accessory navicular on p. 212.)

FOOT AND TOE TRAUMA

Always examine distal neurovascular status.

Fractures and dislocations not discussed below are considered beyond the scope of this book. Consult an orthopedics text or refer the patient for management.

Foot Sprain (*Uncommon*)

Diagnosis

The mechanism of injury is usually excessive plantar flexion or dorsiflexion of the toes and forefoot, injuring the dorsal or plantar soft tissues, respectively. X ray is normal.

Treatment

Use *elevation* and *ice* acutely.

Advise *compression* by *taping* (Figure 18-14) and limited weight bearing, with resumption of activity as tolerated.

Prolonged pain is usually due to a posttraumatic tendonitis, which may be treated with oral antiinflammatory medications or local injection.

Strips are continuous around
bottom of foot.

Fig. 18-14.   Taping a foot. Note that the strips of tape about half overlap each other, and that the foot is not circumferentially constricted.

Fracture of the Base of the Fifth Metatarsal (Jones Fracture) (*Common*)

Fractures of the other metatarsals often require reduction and are not included here.

Diagnosis

The patient reports a sudden inversion stress with the foot in plantar flexion. Tenderness is found over the base of the fifth metatarsal.

X ray confirms the diagnosis. Note that a small accessory ossicle is sometimes present at this location, and should not be confused with a fracture. The fracture line is usually perpendicular to the long axis of the metatarsal, the ossicle usually parallel; the line separating the ossicle from the metatarsal never enters the joint at the base of the metatarsal.

Treatment

Only *symptomatic treatment* is necessary (elevation, ice and compression acutely, and usually a short-leg walking cast applied once the swelling has subsided, left on for 3 to 6 weeks).

However, cases with gross displacement of a large fragment are best evaluated by an orthopedist.

Persistent pain is usually secondary to a posttraumatic peroneal tendonitis, discussed in Chapter 17.

Fracture of the Toes (*Common*)

Diagnosis

The history is one of stubbing the toe or dropping an object onto it.

Treatment

Fractures of the *proximal phalanx* of the great toe or fractures of the distal phalanx that *extend into the interphalangeal joint* should be *referred* to an orthopedist for management.

Fractures of the tuft of the great toe and fractures of the other toes are treated *symptomatically* by *taping* them to an adjacent toe (Figure 18-15) and the use of a roomy or open shoe. A stiff sole is helpful in fractures of the great toe.

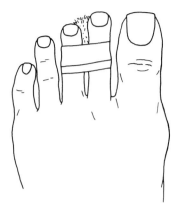

Fig. 18-15. Taping toes. Tape to the adjacent toe (toward the middle toe). Put some lamb's wool or cotton between the toes.

Fractures of the Sesamoid of the Great Toe *(Rare)*

Diagnosis

Usually there is a history of a forceful landing on the ball of the foot.
This must be differentiated on X ray from the smooth edges of a bipartite sesamoid.

Treatment

Use *ice, compression,* and *elevation* acutely.
Then advise *limited weight-bearing* with progression as tolerated.
A metatarsal pad (Figure 18-8) may be helpful.
If pain persists, referral for excision may become necessary.

Dislocation of the IP Joint of a Toe *(Rare)*

Diagnosis

Usually there is dorsal dislocation of the middle phalanx on the proximal phalanx.

Treatment

Reduction is usually easy without anesthesia (Figure 18-16).
A roomy or open shoe and taping (Figure 18-15) are used symptomatically.

Other hand pulls
axially *gently*

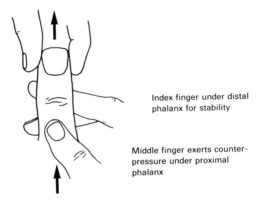

Index finger under distal
phalanx for stability

Middle finger exerts counterpressure under proximal
phalanx

Thumb pushes distally
and downward

Fig. 18-16.   Reducing a dislocated toe.

Patient Handout 18-1

What to Look for in a Shoe

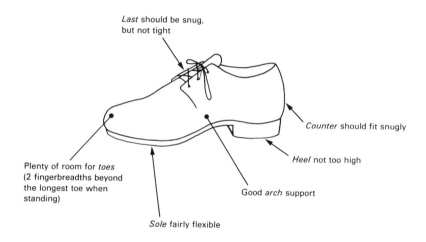

Last should be snug, but not tight

Counter should fit snugly

Plenty of room for *toes* (2 fingerbreadths beyond the longest toe when standing)

Heel not too high

Good *arch* support

Sole fairly flexible

There should be no pressure spots on any part of your foot!

Patient Handout 18-2

Exercises for Chronic Foot Strain

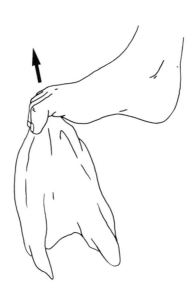

With your foot bare, try to pick up fairly thick and heavy cloth (like toweling) with your toes and hold it for a count of five. Then let go. Do ten repetitions twice a day.

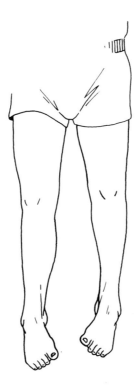

For 10 minutes every day, try to walk around on the outside of your bare feet. Don't do this one if you've got weak ankles, and be careful only to walk barefoot where it is safe. A flat carpeted area is best.

Doing these exercises will help maintain your arch and prevent more problems later.

Patient Handout 18-3

Taking Care of Calluses and Corns

Calluses and corns usually develop at sites of excessive pressure, either from shoes or from one toe against another. So to prevent them from occurring or worsening:

Wear shoes that don't put pressure on the sore area.

Use a donut-shaped pad to take pressure off the sore area (*Note:* the hole goes *over* the corn or callus, the donut *around* it).

If the problem is between the toes, use a moleskin pad or some lamb's wool to avoid the pressure.

If these measures fail to relieve the problem, and the callus or corn is *painful,* you can do the following:

Soak your foot in warm water.

Then, for calluses, *scrape* it down with a clean blade (never, never cut) or with a pumice stone (available in drug stores).

For corns, or for calluses that don't respond to scraping, get a keratolytic agent at the drugstore (something with *salicylic acid* in it) and use it as directed.

19

Pediatric Problems

Musculoskeletal problems in the pediatric age group, aside from trauma, are primarily developmental in nature. Though many of these conditions are self-correcting and require only reassurance, others have great potential for causing lifelong disability if not dealt with aggressively and properly. Only rarely are the young, flexible tissues of the child's musculoskeletal system affected by the degenerative and overuse syndromes that so plague the adult population.

Complex congenital syndromes and bone disease are not dealt with here; the reader is referred to a standard pediatric text for a discussion of these conditions.

Several basic points must never be forgotten when dealing with musculo-skeletal complaints in children:

1. Bony tenderness or a complaint of joint or bone pain in a child should be considered to be a sign of *infection, neoplasm, or serious occult trauma until provem otherwise*. An X ray must be performed, and sometimes a bone scan as well.

2. A toxic infant with a fever without obvious cause should always bring to mind the possibility of a *septic joint* (especially hip).

3. A *hemarthrosis* is a possible cause of joint pain and swelling in children that must be considered because of its grave implication of underlying *blood disease or serious joint trauma*.

4. Injuries involving indirect forces will usually occur at the weakest link in the musculoskeletal unit. In the child this is the epiphyseal plate, i.e., the

Table 19-I   Salter's Classification of Epiphyseal Injury

Salter type	Injury
I	Transverse fracture of only the growth plate
II	Fracture through the growth plate and into the metaphysis
III	Fracture through the growth plate and through the epiphysis
IV	Fracture through the epiphysis, growth plate and metaphysis
V	Impaction

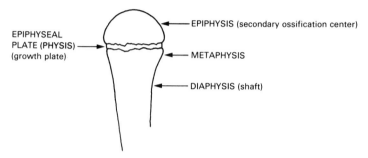

growth plate of bone, or sometimes the bone itself.* Therefore, never make the diagnosis of a sprain, a strain, or tendonitis in a child without carefully ruling out an *epiphyseal plate injury,* which may not be visible on X ray and may manifest itself only as point tenderness at the growth plate. These injuries, if not treated properly, can lead to serious problems indeed.

5. For the same reason, *comparison X ray views of the uninjured limb* are essential.

6. Multiple injuries or those which do not quite fit with the history given must always bring to mind the possibility of a *battered child.*

One more thing: by far the most important intervention most primary care providers will ever make in caring for the musculoskeletal system of (not to mention the rest of) a child is ensuring the use of proper restraints (infant seats, car seats, seat and shoulder belts) while riding in automobiles.

---

*As the child grows into adolescence, the muscle belly and musculotendonous junction become the most vulnerable sites; and in adulthood, it is the tendons themselves or their insertions into the periosteum that absorb the brunt of most indirect injury.

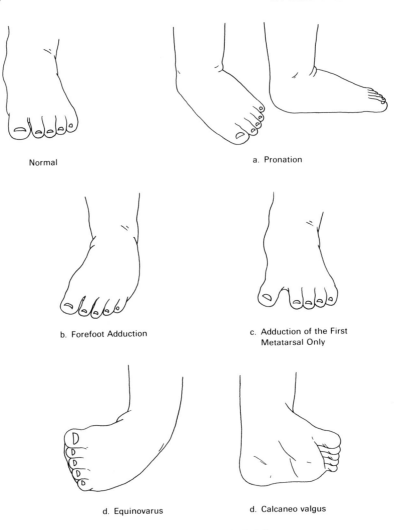

Fig. 19-1. Foot deformities in infants.

CONGENITAL FOOT PROBLEMS

Examination

If all the borders of the foot are straight, the foot is normal. (Note that the plantar fat pad makes the newborn's foot seem flat, but if it is not pronated (Figure 19-1a), it is normal.)

If there is adduction of the forefoot (Figure 19-1b), see if the deformity is passively overcorrectable. If so, see METATARSUS ADDUCTUS, below. If passive overcorrection is not possible, the infant should be referred to the orthopedist.

If there is adduction of only the first metatarsal (Figure 19-1c), see METATARSUS PRIMUS VARUS, below.

If the foot is in equinovarus (Figure 19-1d) or calcaneovalgus (Figure 19-1e), see if the deformity can be passively *over*corrected. If so, no treatment is necessary, but the flexibility of the foot should be rechecked periodically until the deformity corrects. If the deformity cannot be passively *over*corrected, the child has a CLUBFOOT and must be referred to a specialist immediately, as any delay in proper treatment worsens the prognosis.

If the foot is pronated (Figure 19-1a) and cannot be overcorrected by passive manipulation, the patient should be managed by an orthopedist. (Also see the discussion of foot pronation starting on p. 268.)

Metatarsus Adductus (*Common*)

Diagnosis

The lateral border of the foot is convex and the medial border concave, with the head of the talus palpable on the medial side (Figure 19-1b).

The forefoot will flare laterally when the infant's sole is stimulated, and the deformity is passively overcorrectable. (If not, the patient has a form of clubfoot and should be managed by the specialist.)

Pathophysiology

The condition may be secondary to *in utero* positioning.

Treatment

If the abnormality is very mild, no treatment is necessary.

In more obvious deformities, some experts advocate the use of *reverse shoes* (wearing the right shoe on the left foot and vice versa) and *passive stretching* of the forefoot in a valgus direction by the parent several times a day (see Patient Handout 19-1). However, there is a growing belief that no treatment is needed even in these cases. (To reiterate, if the forefoot cannot be passively overcorrected, treatment *is* required and referral to an orthopedist recommended.)

Metatarsus Primus Varus (*Uncommon*)

Diagnosis

The lateral border of the foot is straight, but the first metatarsal points inward, and there is increased space between the first and second metatarsals (Figure 19-1c).

Treatment

None is necessary. However, the parent should be advised never to allow the child to wear shoes with narrow toeboxes, since this will accentuate the deformity and hasten the development of a bunion later in life.

CONGENITAL HIP DYSPLASIA

Evaluation

Ortolani's test (Figure 19-2) should be performed on all infants at birth, and must be repeated at six weeks and three months of age. If the test is positive in the newborn, but abduction of the hip to 90° is possible and there is no other evidence of fixed dislocation (see the following paragraphs), see CONGENITAL SUBLUXABILITY OF THE HIP, below. If the test is positive three weeks or more after birth, the child should be referred to the specialist for management.

Check for a full 90° of passive abduction of each hip in the flexed position (Figure 19-2). This examination should be done both in the immediate postnatal period and at every well-child visit in the first year. If at any time full passive abduction is lacking, the patient should be referred to the orthopedist.

If the Ortolani maneuver is positive, check for shortening on the affected side (by flexing the hips and noting a discrepancy in the level of the flexed knee), for flattening of the involved buttock with the child prone, or for telescoping of the hip (i.e., the ability of the examiner to move the extended leg cranially and caudally relative to the trunk). These findings are quite rare, but if present imply a *dislocated* hip and mandate immediate orthopedic referral.

Congenital Subluxability of the Hip (*Uncommon*)

Diagnosis

A click is found on performing the Ortolani maneuver (Figure 19-2) in the newborn.

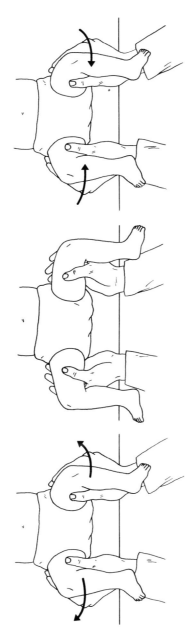

Fig. 19-2. Ortolani's Test and passive hip abduction. Hold one leg in each hand, with both hips and both knees flexed to 90°. Now passively abduct both hips to 90° and return to straight up. An audible or palpable click indicates subluxability of the hip. Lack of a full 90° of passive abduction at any time in the first year mandates referral to an orthopedist.

There is full passive abduction of the hip, and no shortening, telescoping, or flattening of the buttock.

Pathophysiology

Intrauterine malpositioning probably plays a role in some cases by limiting the normal motion necessary for proper joint development.

Genetic factors are probably operative as well, as girls with the condition outnumber boys by four to one, and there are different concordance rates in monozygotic and dizygotic twins.

Maternal hormones may have an effect on joint development.

Treatment

The newborn with a congenitally subluxable hip may be treated with *triple diapering* to maintain a frog-leg position. (The idea is to keep the femoral head firmly seated in the acetabulum until the lax ligaments tighten up a bit.)

Recheck the baby at three weeks of age; if the click persists the infant should be referred to the specialist for further treatment.

If there is limitation of abduction or a sign of fixed dislocation at any time, *referral* is mandatory.

THE PRONATED FOOT IN INFANTS AND CHILDREN

Evaluation

The normal infant foot looks flat because of the presence of a fat pad under the longitudinal arch. If examination reveals a flexible foot, this appearance is of no significance.

If in the newborn infant there is limitation of passive plantar flexion and inversion of the forefoot, there will often be a bony prominence on the medial aspect of the foot. This may be due to any of a number of underlying congenital malformations, and these infants should be referred to the specialist for management.

If pronation of the foot (Figure 19-1a) is noticed when the child begins to bear weight, check the following:

Is the achilles tendon too short? Check by seeing if the foot can be passively dorsiflexed past neutral. If not, the foot must pronate just to allow the heel to touch the ground during walking. These patients should be referred to the orthopedist.

Does the child have an intoeing gait which he is trying to compensate for? If so, he may be consciously forcing his foot into pronation. Check to see if the foot assumes the normal posture when not bearing weight and whether there is good flexibility with passive movement. If so, no treatment is necessary except education.

If the pronation deformity does not disappear when not bearing weight and cannot be passively overcorrected, the child has a RIGID PRONATED FOOT, which should be managed by the specialist.

If the pronation disappears when the child is not bearing weight and can be overcorrected by passive manipulation, the child has a FLEXIBLE PRONATED FOOT.

Flexible Pronated (Flat) Foot (*Common*)

Diagnosis

Pronation can be overcorrected passively by the examiner and disappears when not bearing weight.
A short heel cord is ruled out by normal passive dorsiflexion of the ankle.

Pathophysiology

In the first year or so of weight-bearing, pronation may simply be secondary to the wide-based gait necessary for balance.
If flexible pronation persists beyond that age, it is probably a consequence of ligamentous laxity.

Treatment

Before three years of age, no treatment for the flexible flat foot is necessary.
Beyond that age, if the deformity is mild, no treatment is required either.
If the deformity is fairly severe, or if the child complains of foot pain after activity (and no other etiology is found; see the next section), *exercises* should be prescribed as shown in Patient Handout 18-2. The child should be encouraged to do as much running as he can and to go barefoot when possible. (Again, be sure you are not dealing with a rigid flat foot.)

FOOT PAIN IN CHILDREN AND ADOLESCENTS

Evaluation

Be sure that the problem is not simply poorly fitting shoes creating areas of excessive pressure.

As in any musculoskeletal pain syndrome in children, X rays must be performed to rule out infection, neoplasm, or an occult fracture.

If the child complains of diffuse pain that is worse after activity, and examination reveals excessive pronation but a flexible foot, the cause is probably CHRONIC FOOT STRAIN; see the discussion of the treatment of the mobile pronated foot, above.

Pain and tenderness in the area of the second metatarsal head in a preteenager or adolescent is likely due to FREIBERG'S DISEASE.

Pain and tenderness in the medial foot in a prepubescent or pubescent child suggests KOHLER'S DISEASE.

Freiberg's Disease (*Rare*)

Diagnosis

Pain and tenderness are found under the second metatarsal head (Figure 18-9) in an adolescent.

The X ray may not show changes for several weeks.

Pathophysiology

Unknown. Falls into the category of the osteochondroses.

Treatment

Symptomatic measures (decreased activity, analgesics, a donut-shaped pad under the sore area) and usually all that is necessary. The problem resolves spontaneously in a few weeks to a few months.

Kohler's Disease (*Rare*)

Diagnosis

Pain, swelling and tenderness of the tarsal navicular bone are found in a prepubertal or pubertal child.

The X ray may not show anything early in the course.

Pathophysiology

Unknown. An osteochondrosis.

Treatment

Advise the patient re *symptomatic measures* (decreased activity, analgesics, open shoes, or padding over the painful area).

The condition resolves of its own accord over several weeks to several months. Severe cases should be referred for consideration of casting.

HEEL CORD PAIN IN CHILDREN

Unless it is simply a pressure phenomenon from ill-fitting shoes, this is invariably due to SEVER'S DISEASE.

Sever's Disease (*Uncommon*)

Diagnosis

Pain and tenderness occur at the achilles tendon insertion in an adolescent or preadolescent.

The X ray may or may not show fragmentation at the calcaneal apophysis.

Pathophysiology

This is a traction-type injury at the apophysis where the achilles tendon attaches; analogous to Osgood-Schlatter's Disease in the knee.

Treatment

A $\frac{1}{4}''$ *heel lift* in both shoes will be helpful.

Decrease activity to achieve a level of pain acceptable to the patient.

Aspirin and *heat* may be used as needed.

The natural course of the condition is for it to disappear completely when the apophysis closes.

INTOEING GAIT

Evaluation

Observe the child and estimate the angle of gait, i.e., the angle the feet make with the line of progression. Normal is neutral to 30° of outtoeing.

Examine the foot for a METATARSUS ADDUCTUS deformity (Figure 19-1b). This condition was discussed earlier in this chapter.

The degree of TIBIAL TORSION can be evaluated by measuring the thigh-foot angle (Figure 19-3a). At birth, this is usually about zero degrees, though some infants are born with residual internal torsion. The tibia rotates outward approximately 20° by the time the child is walking. If abnormal, see below.

Measure hip rotation as shown in Figure 19-3b. The sum of external and internal rotation is usually about 100°. If internal rotation is greater than 70°, excessive FEMORAL ANTEVERSION is present; if more than 80°, the abnormality is considered severe. (Femoral anteversion is the angle made between the femoral neck and a line between the distal femoral condyles. The normal child has 40° or so of femoral anteversion at birth, decreasing to 10° by about age eight.) See the discussion below.

Internal Tibial Torsion (*Common*)

Diagnosis

As explained above.

Treatment

Although some authorities still recommend the use of a night brace, the trend is toward advising no treatment at all, except perhaps the avoidance of habitual sleeping and sitting positions which reinforce and possibly perpetuate the deformity.

Excessive Femoral Anteversion (*Uncommon*)

Diagnosis

As above.

Treatment

Since the natural course is for the abnormality to decrease with time, and for compensatory external tibial torsion to occur, no treatment is necessary unless by

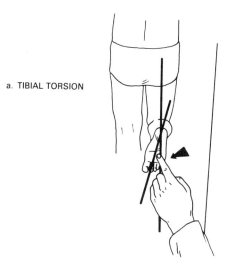

a. TIBIAL TORSION

Estimate the angle between the thigh and foot.

b. FEMORAL VERSION

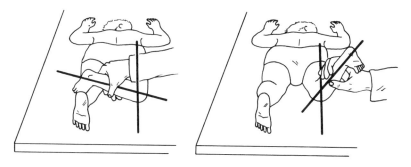

Estimate maximal internal rotation and external rotation (see text).

Fig. 19-3.   Evaluating tibial torsion and femoral version.

age eight there is still a gait abnormality severe enough to warrant the consideration of surgery.

BOWLEG, KNOCKKNEE, AND BACKKNEE

Evaluation and Management

Bowing of the legs (GENU VARUM; see Figure 19-4a) is normal until about
the age of eighteen months, at which time it should begin to correct, and eventu-

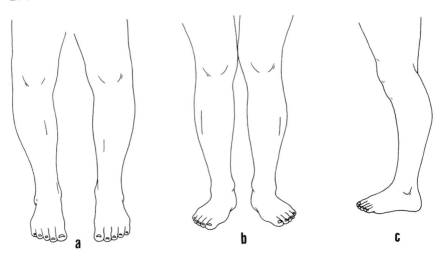

Fig. 19-4.   Genu varum, valgum and recurvatum.

ally may result in a knockknee deformity (GENU VALGUM; see Figure 19-4b), which in turn may normally persist to age five to seven. No treatment is required, except as noted in the following paragraphs.

If bowing does not begin to decrease by age two, or if it is severe or unilateral, an A-P X ray of the legs should be performed to rule out rickets, Blount's Disease (osteochondritis of the medial proximal tibial epiphysis) or other metabolic or developmental conditions.

Similarly, if knockknee is severe (more than 15° or so) or unilateral, or does not begin resolving by age seven, radiographs should be obtained to rule out underlying disease.

Congenital GENU RECURVATUM (Figure 19-4c) is usually due to intrauterine malpositioning and is especially common in breech babies. Tibial subluxation must be ruled out by X ray; if it is, no treatment is necessary and the deformity usually self-corrects in a few weeks. If it persists, however, the infant should be referred to the orthopedist for splinting, etc.

KNEE PAIN IN CHILDREN AND ADOLESCENTS

Evaluation

Never forget that knee pain can be the sole complaint of a child with serious disease of the *hip*. Every complaint of knee discomfort mandates an examination of the hip.

An X ray should always be done to rule out infection, neoplasm, or occult trauma.

Pain and tenderness, often with a lump and often bilateral, over the tibial tubercle in a preteen or adolescent suggests OSGOOD-SCHLATTER'S DISEASE.

Aching pain that is worse with activity (especially climbing hills or stairs), stiffness after sitting with the knee flexed, and painful grating on passive compression and movement of the patella against the femur suggest CHONDROMALACIA PATELLAE. This condition is common in teenagers, especially girls, and is discussed in Chapter 15.

A history of episodes of knee collapse associated with lateral displacement of the kneecap suggests RECURRENT PATELLAR SUBLUXATION, a condition most often found in teenage girls.

A history of collapse and/or vague pain may be found with chondromalacia, but also could be indicative of OSTEOCHONDRITIS DISSECANS or a MENISCAL TEAR. These conditions are discussed in Chapter 15.

For further details about evaluation of the painful, collapsing, swollen, or injured knee, see Chapter 15.

Osgood–Schlatter's Disease (*Common*)

Diagnosis

Tenderness, often with a lump, will be found over the tibial tubercle (Figure 15-4) in a rapidly growing adolescent.

It is most common in athletically inclined boys.

X ray may or may not show fragmentation at the tubercle.

Pathophysiology

This is a traction-type injury on the apophysis where the patellar tendon attaches; some hold that it is due to tendon growth not keeping up with bone growth.

The same process occurring at the other end of the patellar tendon (i.e., the inferior pole of the patella) is called Larsen-Johannson's disease.

Treatment

Decrease activity to achieve a level of pain acceptable to the patient.

Aspirin and *heat* may be used as needed.

The natural course of the condition is for it to disappear completely when the apophysis closes.

Recurrent Patellar Subluxation (*Uncommon*)

(See also the discussion of acute patellar dislocation in Chapter 15.)

Diagnosis

The patient will report episodes of collapse of the knee associated with lateral displacement of the kneecap, often with some swelling afterward.

Pathophysiology

The instability may be due to a congenitally flat femoral condyle or shallow patellar ridge, or to external tibial torsion or previous acute patellar dislocation.

Treatment

Quadriceps strengthening exercises are often very helpful (see Patient Handout 15-2).

If that does not give sufficient stability, referral should be made to an orthopedist for stabilizing procedures or bracing.

HIP PAIN IN CHILDREN AND ADOLESCENTS

Evaluation

A child with hip pain or limitation of motion who has more than a low-grade fever, or who has an elevated WBC count or sedimentation rate, or who appears ill, must be considered to have a septic joint until aspiration proves otherwise. The septic hip is more often found in the infant (where a unilaterally flexed hip, refusal to move the leg, or crying with diapering may be the only clues to the cause of the child's systemic toxicity) but can occur at any age. Rapid destruction of the joint occurs if aggressive therapy is not instituted immediately; STAT consultation with the orthopedist is mandatory.

Ask about other joint involvement to help rule out juvenile rheumatoid arthritis.

Thomas' test (Figure 19-5) can reveal whether there is a flexion contracture of the hip, which is a sign of intrinsic hip disease.

An X ray must be done to rule out osteomyelitis, tumor or occult trauma, as well as to look for signs of the three conditions discussed below.

Keep in mind that hip disease often manifests itself as a complaint of pain in the knee, abdomen or back instead of the more expected groin pain or limp.

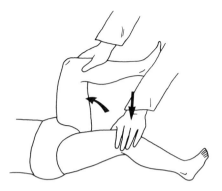

Fig. 19-5. Thomas' test. Flex the opposite hip and knee 90° to flatten the pelvis, then see if you can passively fully extend the hip you are testing. If not, a flexion contracture is present.

Atraumatic hip pain in the child or adolescent, if the conditions mentioned above are ruled out, may be due to one of three diseases, all of which must be considered in the differential diagnosis.

Transient Synovitis of the Hip (*Uncommon*)

Diagnosis

Pain and limp may be of fairly sudden onset, or can come on gradually over one to two weeks.

There can be a low grade fever, but if there is significant elevation of temperature, white cell count or sedimentation rate, or if the child appears toxic or in much pain, a septic arthritis must be ruled out by referring the patient immediately for joint aspiration.

The condition is most commonly seen from ages three to six, but can occur up to age twelve.

On examination, the child will have a toe-walking gait and will hold the hip in flexion.

Range of motion will be limited and painful at the extremes. (In septic arthritis *any* motion will be painful.)

X ray may be normal or only show evidence of a joint effusion.

Pathophysiology

The synovitis may be a result of unrecognized trauma or possibly of a viral infection.

Treatment

The child should be placed at *bedrest*, with *aspirin* used for pain.
Resolution usually occurs over several days. When reexamination reveals pain-free range of motion and loss of the flexion contracture, weight-bearing may be resumed.

All children with this condition must be reexamined and X rayed again after several months to rule out LEGG-PERTHE'S DISEASE (see below) which is eventually found in one-tenth of cases originally diagnosed as transient synovitis.

Legg-Perthe's Disease (Avascular Necrosis of the Femoral Head) (*Uncommon*)

Diagnosis

This can be a cause of hip pain in children three to twelve years of age, in boys more often than in girls.

Groin pain (sometimes referred to the knee or abdomen) and a limp may begin fairly abruptly but more commonly are noted to come on over weeks to months. Symptoms are exacerbated by activity and somewhat relieved by rest.

Examination reveals a flexion contracture and limitation of motion, with pain at the extremes.

X ray is characteristic (showing increased density of the femoral head) but may remain normal for the first couple of months, and so must be repeated at intervals before the condition can be ruled out.

Sickle cell disease should be tested for in black children with this condition.

Pathophysiology

This is an avascular necrosis of the femoral head, the etiology of which is unknown and probably multifactorial. It is often classified with the osteochondroses, but whether it belongs there is a matter of conjecture.

The disease passes through three stages: necrosis, revascularization, and reossification.

Treatment

Proper management is complex and is critical to prevent severe hip problems later in life. The patient should be managed by an orthopedist.

Slipped Capital Femoral Epiphysis (*Uncommon*)

Diagnosis

It is most commonly found in ages ten to seventeen, especially in tall and thin boys or conversely, in quite obese ones. One-fourth are bilateral.

The onset of pain in the groin (or often referred pain to the knee) and a limp may be sudden after trauma, or gradual over weeks to months.

On examination there will be a flexion contracture and also quite marked restriction of internal rotation.

A-P and lateral X rays will usually show the slip, but if they do not and suspicion remains, a true lateral view should be ordered and may be revealing. (The femoral head normally looks like a scoop of ice cream on a cone; in epiphyseal slippage, the ice cream appears to be sliding off the cone.)

Pathophysiology

Hormonal factors are thought to play a role in weakening the epiphyseal plate and allowing the slip posteriorly and inferiorly of the femoral head.

Treatment

This condition should be considered an *emergency* and referred to the orthopedist immediately for pinning.

BACK PAIN IN CHILDREN AND ADOLESCENTS

Evaluation

History and examination are as discussed in Chapter 11. Children can have ACUTE BACK STRAINS though much less commonly than adults. ACUTE LUMBOSACRAL RADICULITIS secondary to a ruptured intervertebral disc is quite rare in childhood. These conditions are covered in Chapter 11.

The possibility of pain *referred* from the kidneys, other retroperitoneal structures, the abdomen, or a hip must be kept in mind, and conditions there must be ruled out by appropriate examination and laboratory tests.

An X ray must be performed in all children with backache, especially to rule out osteomyelitis or tumor. If SPONDYLOLISTHESIS is found, see Chapter 11. Shmorl's nodes are diagnostic of SCHEUERMANN'S DISEASE, which is discussed below. Abnormalities in the curvature of the spine are discussed in the next section.

A child with persistent backache and a normal X ray may well have DISCITIS or early OSTEOMYELITIS. A bone scan will be positive before the X ray shows any changes; an elevated sedimentation rate is almost invariably found with these conditions. See Chapter 11.

Scheuermann's Disease (*Rare*)

Diagnosis

A complaint of back pain is given by an adolescent, and sometimes an increased kyphosis is seen on exam.
Shmorl's nodes are seen on X ray.

Pathophysiology

This is an idiopathic breakdown of the developing vertebral end-plate.
The breakdown allows herniation of disc material into the vertebral body and can lead to an increasing kyphosis.

Treatment

Extension exercises and *postural instruction* are the mainstays. These patients must be observed carefully for the development of kyphosis which can lead to severe back problems later in life. At the first sign of increasing curvature the patient should be referred to a specialist for consideration of bracing, etc.
Otherwise, *rest* and *antiinflammatory medication* should be used as needed.
The disease becomes asymptomatic spontaneously.

ABNORMAL CURVATURE OF THE BACK

Evaluation

A child or adolescent with excessive kyphosis (roundback) should be referred to a specialist for management.
Any lateral curvature (SCOLIOSIS) is abnormal. A *functional* scoliosis is present if the curve disappears on recumbency. It may be due to leg length inequality, muscle contracture about the hip, or splinting secondary to back pain from any other condition. Correction is by treatment of the underlying cause. A *structural* scoliosis remains present even with a change in position, and is a much more serious matter. A screening examination for scoliosis as described below

must be performed yearly on preteens and adolescents. (More and more schools are instituting such programs.)

Structural Scoliosis (*Uncommon*)

Diagnosis

Lateral curvature does not disappear with recumbency.
An asymmetric rib hump or asymmetry of the shoulders, scapulae or lumbar muscle masses can be seen from behind when the patient bends forward with the hands together (Figure 19-6).
Flank creases may be asymmetric.
A plumb line dropped from the center of the cervical spine misses the gluteal cleft. (This test may fail in a compensated double-curve scoliosis.)

Pathophysiology

Most cases are idiopathic and thought to be genetically determined, but specific vertebral malformations or neuromuscular disease may sometimes be responsible.

Treatment

Proper management is essential to avoid possible skeletal, neurologic and cardiorespiratory complications. All children and adolescents with any degree of structural scoliosis noted on examination or X ray must be *referred* to an expert in the field for follow-up.
Once epiphyses have closed, progression is not likely unless the degree of curvature is great; over the age of 25 only patients with more than 25° of scoliosis need be referred. The management of back pain in those with lesser curves is similar to that of other patients with back pain.

WRYNECK IN INFANTS AND CHILDREN

Evaluation and Management

Infants and young children with a *fixed* lean of the head to one side and rotation toward the other side have a CONGENITAL MUSCULAR TORTICOLLIS (unilateral contracture of the sternocleidomastoid muscle). If the deformity is mild and there is no mass palpable in the sternocleidomastoid muscle, advise the parents to do the following:

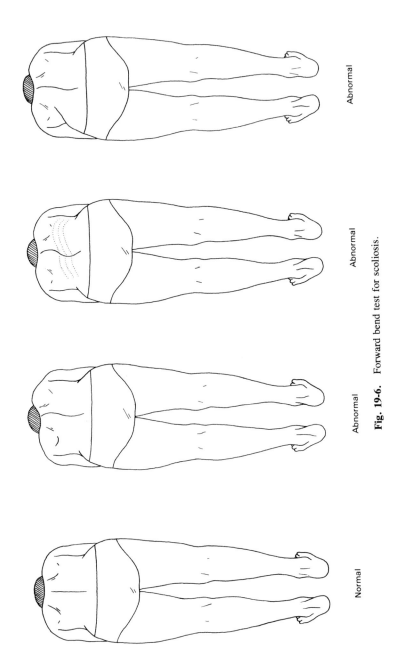

Normal   Abnormal   Abnormal   Abnormal

Fig. 19-6. Forward bend test for scoliosis.

Position the infant for sleep prone, with the head turned to alternate sides.

Position him seated so that he must turn his head to stretch the affected muscle in order to see what is going on.

Gently stretch the tight muscle at each diaper change.

If the deformity is severe or a mass is palpable, the child should be referred to the specialist for treatment. Note that a fair percentage of these children also have congenital hip dysplasia.

Sudden atraumatic onset of pain on one side of the neck with deviation of the neck (ACUTE TORTICOLLIS) is managed as in adults. See Chapter 4.

ELBOW PROBLEMS IN CHILDREN

The pulled elbow in a toddler and elbow pain in the child who does much throwing are the only common afflictions of this joint (aside from fractures) in the pediatric age group.

Radial Head Subluxation (Pulled Elbow, Nursemaid's Elbow) (*Uncommon*)

Diagnosis

The injury usually results from a toddler one to four years of age being pulled or swung by his or her extended arm, as when a parent tries to help the child negotiate a curb.

The child holds his forearm pronated and his elbow flexed.

The X ray is normal.

Pathophysiology

The head of the radius subluxes through the annular ligament, which normally holds it in place.

Treatment

The subluxation is *reduced* by grasping the child's forearm with one hand while pressing the thumb of your other hand over the radial head; the forearm is supinated and the elbow flexed, and the subluxation reduces with a palpable snap. After a few moments the child is pain free and no further treatment is necessary.

With recurrent subluxations, the use of a *sling* for a few weeks is helpful.

Little League Elbow (*Uncommon*)

Diagnosis and Management

In a child that overuses the elbow as in throwing, two types of conditions are commonly seen.

One is a traction apophysitis of the medial epicondyle (see Figure 7-2). Throwing must be curtailed or eliminated until symptoms resolve or the injury may progress to a true avulsion fracture, with subsequent growth disturbance and deformity.

The other problem is an avascular necrosis of the capitellum of the humerus called *Panner's Disease*, which is probably secondary to repetitive compression force. X rays may be normal early. Throwing must be completely stopped and the patient referred to an orthopedist.

Possibly a sequel of the latter is *Osteochondritis of the Elbow*, in which a chip of articular cartilage with or without a piece of attached bone breaks off and becomes a loose body in the joint, leading to recurrent pain and swelling and sometimes locking. Tomograms may be necessary to make this diagnosis. If symptoms warrant, the patient should be referred to the orthopedist.

Note that the development of a flexion contracture is a nonspecific sign of significant elbow problems; it mandates the complete discontinuance of throwing as well as referral for proper management.

MUSCULOSKELETAL TRAUMA IN CHILDREN

We here reiterate three of the principles discussed in the opening of this chapter and add another.

Any periarticular trauma in a child must be considered to be a possible epiphyseal plate injury. These can have serious consequences, and point tenderness over an epiphysis may be the only finding. (X rays can be normal.)

Comparison views of the uninjured side should always be taken.

Multiple or repeated trauma or a history not quite consistent with the physical findings could mean the *Battered Child Syndrome*. This is a potentially life-threatening situation. Failure to make the diagnosis and act appropriately on it, whether due to oversight or to fear of being wrong and offending the parents, is a mistake that must be avoided.

Always check neurovascular status distal to any injury.

Tuft fractures of the distal phalanges and fractures of the small toes are managed as in adults (see Chapters 9 and 16). TORUS FRACTURES of the forearm and CLAVICULAR FRACTURES are discussed below. Dislocation of the radial head was discussed above. All other fractures and dislocations in children are beyond the scope of this book; the reader is advised to refer the patient to an orthopedist or to consult an orthopedic text or manual. Bruises and myositis ossificans are discussed in Chapter 21.

Torus Fractures of the Distal Radius (*Common*)

Diagnosis

X ray reveals a buckling of one or both sides of the cortex of the distal radius, usually after a fall on the outstretched hand.

If there is tenderness over an epiphysis, or if on any view there is a discontinuity of cortex, the patient has more than a torus fracture and treatment is more complex.

Treatment

Protective splinting for three to four weeks and a follow-up X ray to assure healing are the only treatments necessary for a torus fracture.

Clavicular Fracture in a Child (*Common*)

Diagnosis

Careful examination must be made to rule out neurovascular compromise.

The younger the child, the more likely the fracture is to be of the "greenstick" type (with one side of the cortex intact) rather than a complete break.

Treatment

If there is more than 30° of angulation, direct pressure over the injury site should be used to decrease the deformity.

A figure-of-eight bandage should be applied and the parents taught to retighten it daily. Two to three weeks is usually sufficient.

SYSTEMIC ILLNESSES

Juvenile Rheumatoid Arthritis (*Rare*)

Clinical Features

There are three fairly distinct types:

1. *Still's disease* is an illness in which systemic manifestations often over-shadow joint complaints. There are daily fever spikes with in-between returns to normal temperature, a faint salmon-colored rash, lymphadenopathy, and some-times hepatosplenomegaly. The white cell count and sedimentation rate are increased.

2. *Pauciarticular arthritis* involves a few large joints and is usually seen in younger children. This is the most common type. About one-fourth of these patients will develop potentially serious eye disease.

3. *Polyarticular arthritis* is seen with greater frequency as children grow older, and is almost certainly the same disease as adult rheumatoid arthritis.

Pathophysiology

The etiology of each of the three types is unknown. Joint pathology is analo-gous to that of adult rheumatoid arthritis.

Treatment

Children with pauciarticular disease must be referred for ophthalmologic ex-amination at least twice yearly, even if asymptomatic.

Treatment modalities are analogous to those used in adult rheumatoid arthritis. Growth retardation can result near severely affected joints.

Rheumatic Fever (*Rare*)

Clinical Features

The arthritis of rheumatic fever characteristically is migratory and involves at least several joints; the originally involved joints become asymptomatic as other joints become inflamed. Large and medium-sized joints are most commonly affected.

Jones' criteria for the diagnosis of rheumatic fever are listed below. Evidence of preceding streptococcal infection (a positive throat culture or increased antistreptococcal antibodies) plus two major (or one major and two minor) criteria indicates the probable presence of rheumatic fever.

Major	Minor
Carditis	History of previous rheumatic fever
Polyarthritis	Evidence of rheumatic heart disease
Chorea	Arthralgias
Erythema marginatum	Fever
Subcutaneous nodules	Increased WBC count and ESR
	ECG changes

Pathophysiology

The mechanism by which streptococcal infection sometimes leads to rheumatic fever remains unknown.

Treatment

Antiinflammatory medications, antibiotics, rest, and *specific treatment for complications* (such as congestive heart failure) are used. The reader is referred to textbooks of pediatrics or internal medicine for details.

Patient Handout 19-1

Stretching Your Infant's Foot

Your child has a mild foot deformity called *metatarsus adductus*. This usually gets better with time; you can help speed up the process by stretching the foot as shown below ten times at each diaper change.

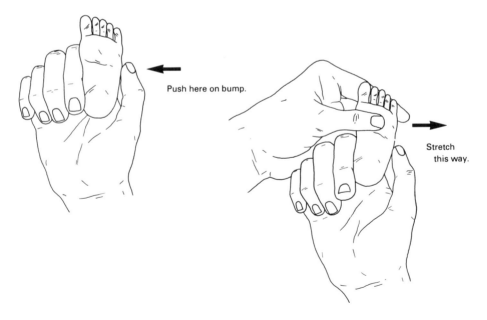

Push here on bump.

Stretch this way.

Patient Handout 19-2

Your Baby's or Child's Shoes

Unless your child has a foot deformity for which the doctor has made specific recommendations, he or she needs shoes for only one reason: protection from injury. Your child should go barefoot whenever it is safe; when it isn't, dress him or her in shoes that are large enough, soft, flexible and well ventilated. The sole should be flat; high tops are helpful only in making the shoe hard to remove. Remember, a rigid shoe will slow the development of your child's foot muscles.

20

Running Injuries

TYPES OF PROBLEMS

More and more Americans are running longer and longer distances, and many are paying a musculoskeletal price for their psychological and cardiovascular benefits. Those who see doctors not attuned to the biomechanics or the psychology of running are likely to be told simply to stop or cut down on their avocation, or possibly offered medications or injections; they are not likely to be offered advice about their running style, training method, or shoe that can relieve the problem or at least prevent its recurrence when natural healing has taken place and they are ready to resume their activity. For many runners this is unacceptable, and they will seek advice from podiatrists, other doctors, or sports medicine specialists, or sometimes, unfortunately, from unqualified friends or from quacks. This chapter is an effort to change that, so that primary physicians can give sound and acceptable advice to their running patients based on an understanding of the pathophysiology of common running injuries.

While the diagnosis and treatment of these problems are not overly complicated, they are not blatantly simple either. Even some writers of articles and books on running medicine commonly lump disparate problems under titles like "shinsplints" or "runner's knee," for example, and then prescribe the same treatment for all runners with leg pain or with knee pain. This must be recognized as an oversimplification that will often lead to suboptimal therapeutic advice.

There are three general categories into which lower-extremity problems in runners can be grouped (some conditions may fit into more than one category):

1. *Weight transmission problems.* These occur because in running (as opposed to walking) there is a phase in which both feet are off the ground. When the heel then strikes the surface, several times body weight must be absorbed somewhere, and over a period of time several types of injury can result. This tends to especially be a problem in runners with rigid (usually high-arched) feet, or those increasing their mileage too quickly.

2. *Biomechanical problems.* If because of terrain, shoe, running style or anatomy, excessive stretch is placed on a structure, damage can result. The injury can actually occur proximal to the area where the biomechanical anomaly is occurring, as intermediate structures partially compensate for the strain but in turn stress other parts of the leg.

3. Runners are also subject to the same diseases and conditions as others. A good biomechanical approach to their lower extremity complaints must not lead to overlooking such conditions as deep vein thrombosis, meniscal tears in the knee, and osteomyelitis, for example.

TREATMENT PRINCIPLES

General

1. Most running injuries occur with a change in training routine: most often new shoes, new terrain, or a too-rapid increase in mileage. Increase in distance should always be slow and gradual.

2. It is now widely accepted that *alternating days of heavy training with days of light training* leads to less microtrauma to musculoskeletal tissues. Most probably this is simply a manifestation of the time necessary for these tissues to at least partially recover before being retraumatized. While this is a good rule to follow for all runners, it is especially important when the patient is working his or her way back from an injury.

3. For some injuries (such as a tibial stress fracture) *a temporary cessation* of running, with slow resumption after a period of time, will be necessary. But for

most, just a *decrease in mileage* to a point where pain is avoided, along with correction of training errors and/or biomechanical factors, and then *slow reincrease in distance,* is more appropriate and acceptable to the patient. If running must be temporarily discontinued, the substitution of *other aerobic activity* such as swimming or bicycling makes the prescription easier to swallow.

4. *Good shoes* are extremely important. Runners with low arched flexible feet should wear shoes that provide especially good support and resistance to pronation, while the high-arched rigid foot needs extra help in shock absorption. See Patient Handout 20-1.

5. The application of an *ice* pack to the sore area for several minutes before and after running may be of value.

6. In more severe degrees of injury of either the weight-transmission or biomechanical types, a course of oral *antiinflammatory medication* can often be useful in shortening the duration of disability.

Weight Transmission Problems

1. *Shoes* that have *good shock absorption* characteristics must be used. See Patient Handout 20-1.

2. The patient should be advised to run only on *softer surfaces* such as grass or dirt or a synthetic track and to avoid concrete and asphalt. (But too-soft surfaces such as sand can lead to problems of their own and should be avoided.)

3. Running should be done on *flat terrain.* Coming down hills increases transmitted force significantly.

4. If possible, the runner should *increase* his pace. The widespread belief that slowing down is indicated is probably erroneous; each heel strike transmits less force at higher speed. This makes sense if you think about it: the fast runner is gliding over the track, while a slow one crashes his weight into the ground with every step.

Table 20-I   Measures to Decrease Transmitted Weight

1. A shoe with especially good shock absorption characteristics should be chosen.

2. Running should be done on fairly soft surfaces.

3. Hills should be avoided.

4. A slightly faster pace (if feasible) can be beneficial.

Biomechanical Problems

1. Whether a certain running style, shoe or anatomical aberration needs to be corrected depends only on whether it is causing symptoms. It makes no sense to devote a detailed examination and specialized testing to detect mild deviations from the anatomical or biomechanical norm if the body has compensated for them. As a matter of fact, the correction of compensated minor aberrations can itself lead to injury. Conversely, if a certain structure is becoming painful with running, the specific features of style, shoe, terrain or anatomy that may be responsible should be looked for and corrected. (Specific possible anomalies responsible for a given complaint are discussed in the section on that syndrome.)

2. Examine the shoe specifically for localized wear or for a repair done by the runner that overcorrects the wear. Either of these can lead to subtle but nevertheless injurious alterations in previously well-compensated biomechanics. In addition, shoe-wear patterns can be a valuable diagnostic clue to abnormal biomechanics and poor running style. See Figure 20-1.

3. One biomechanical anomaly is so widespread and so commonly leads to problems that it deserves special mention here: *excessive foot pronation*. Normally, immediately after heel strike, weight is transferred along the outside border of the foot; at this moment the foot is in mild supination. As more of the foot comes into contact with the ground, the foot moves into neutral and then into pronation, allowing the foot to conform to the surface. Weight moves from the lateral side across the transverse arch toward the ball of the foot; pushoff is then accomplished by the large toe as the foot momentarily returns to supination to act as a rigid lever. If the foot goes into pronation too soon or too far, a variety of problems can result in the arch, ankle, shin and knee, as noted in the discussion of specific syndromes below.* (These feet tend to transfer weight in more of a straight line from the heel to the middle of the transverse arch.) *Note:* Sprinters, some novice runners, and some very experienced distance runners actually strike the ground more toward the front of the foot.

Running on an *inclined surface* (such as a beach or the side of a road), with one leg higher then the other, causes excessive pronation in the higher foot and too much supination in the lower one. The patient should be advised to avoid this type of terrain. Runners who go around an oval or round track *always in the same direction* will tend to pronate the outside foot excessively; alternating directions or avoiding this type of track altogether is indicated.

Shoes that control hyperpronation by holding the heel snugly and supporting

---

*It is also conceivable that severe hyperpronation, through malalignment of the subtalar joint, may contribute to poor shock absorption and thus to the weight transmission problems discussed previously.

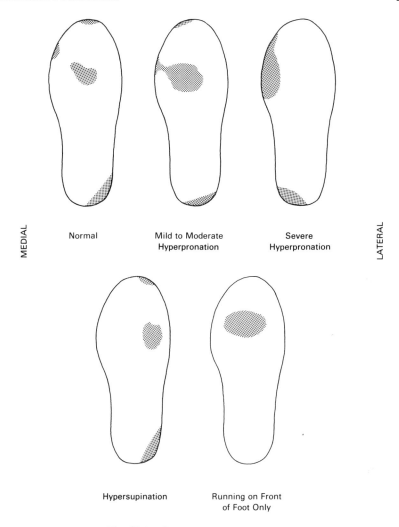

MEDIAL   Normal   Mild to Moderate   Severe   LATERAL
Hyperpronation   Hyperpronation

Hypersupination   Running on Front
of Foot Only

Fig. 20-1.   Shoe wear patterns in runners.

the longitudinal arch should be chosen. If simply changing shoes or inserting a soft noncustomized arch support is ineffective, the runner may have to resort to a considerably more expensive customized *orthosis* (''orthotic''). These can be obtained from orthotists and many podiatrists. Finally, a *medial heel wedge* can be placed on the shoe. (Be careful with this, because though quite effective at correcting pronation, they can predispose to *ankle sprains*.)

Table 20-II   Measures to Decrease Excessive Foot Pronation

1. Running on banked surfaces (with one foot higher than the other) must be avoided.
2. Running around small tracks (where the runner is in a turn for a significant part of the total time) likewise should be discouraged.
3. The shoe should cradle the heel snugly.
4. Additional support for the longitudinal arch should be provided, either in the construction of the shoe, with a soft, noncustomized arch support insert, or if sufficient relief isn't obtained, with a customized orthotic.
5. The medial heel of the running shoe may be built up externally a bit, or a medial heel wedge can be incorporated into an orthotic. (Be careful with these because of possible predisposition to ankle sprains.)

SKIN PROBLEMS

For a discussion of calluses and corns, see Chapter 18.

Blisters (*Extremely Common*)

Diagnosis

By inspection. They are usually related to new shoes or marked increases in mileage.

Pathophysiology

Localized shearing stresses lead to separation of the skin layers, with fluid extravasation between them.

Treatment

Large blisters should be *punctured* (not denuded) and *drained* using sterile technique.

Open blisters must be conscientiously kept *clean* to prevent infection.

Small or incipient blisters should be covered with *adhesive tape* to protect them from shearing forces.

Well-fitting shoes worn without socks will help prevent the development of blisters.

TOE PROBLEMS

Runner's Toe (*Common*)

Diagnosis

There will be inflammation and sometimes ecchymosis, especially near the nail bed.

Pathophysiology

Most commonly it is due to shoes that are too short with subsequent axial pressure on the nails.

Sometimes the problem can be a consequence of a habit of running with the toes curled, with the tips being pounded into the bottom of the shoe; this running style can be a consequence of wearing shoes that are too *big* or of coming downhill.

Treatment

Be sure *shoes* are of the proper length; one to two cm beyond the tip of the toes when standing is about right.

Nails obviously must be kept *well-trimmed*.

If the patient persists in running with his toes curled, temporary placement of a pad over the dorsal PIP joints of the involved toes can help cure him of the habit.

PAIN IN THE BALL OF THE FOOT

See the discussion in Chapter 18. In the treatment of these conditions, be careful in stiffening the sole and lowering the heel, because the altered biomechanics can lead to achilles tendon and calf problems.

FOREFOOT PAIN

Evaluation

If there is localized tenderness over the dorsal aspect of one or more metatarsal bones, the patient probably has a METATARSAL STRESS FRACTURE (see below).

Tenderness maximal under the second (sometimes the third) metatarsal head probably means METATARSALGIA (see below). But in a preteen or adolescent consider FREIBERG'S DISEASE (Chapter 19).

Tenderness between the third and fourth metatarsal heads (sometimes the second and third), with pain on lateral compression of the forefoot, indicates a MORTON'S NEUROMA. See Chapter 18.

Metatarsal Stress Fracture (*Common*)

Diagnosis

Localized tenderness is found dorsally over one or more metatarsal shafts, usually in a runner who is increasing his mileage or is running on hillier terrain or a harder surface.

The X ray may be normal for the first couple of weeks, but the diagnosis can be made by bone scan before X ray changes are apparent. (A scan is necessary only if the patient is unwilling to rely on your clinical judgment, as runners often are.)

Pathophysiology

This is a problem of excessive transmitted stress. (See Chapter 1 for a discussion of the pathophysiology of stress fractures.) Running "on the toes" or uphill can be contributory.

Treatment

If very painful, *crutches* may be used for a few days, or a short leg walking cast can be applied for a week or two, but these measures are for symptomatic relief only.

Running should not be resumed for three to four weeks, and then mileage slowly increased using a heavy/light alternating day schedule.

When running is resumed, it should be done only on flat, soft surfaces, and the shoe should have good shock absorption characteristics throughout (not just the heel). *Running "on the toes" should be discouraged.*

Metatarsalgia (*Common*)

Diagnosis

Tenderness, and often callus formation as well, is found under the second (sometimes the third) metatarsal head, usually in an older runner (Figure 20-2).

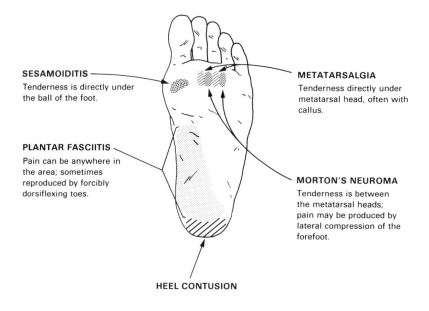

Fg. 20-2. Pain in the bottom of the foot.

Pathophysiology

Hyperpronation leads to weight transmission more directly from heel to transverse arch, with excessive pressure resulting there.

Treatment

A soft or hard *arch support* should be used, and the heel counter must be snug.

If this and a *temporary decrease* in running do not bring relief, consider the use of a *metatarsal pad* (Figure 18-8) and/or a course of *antiinflammatory medication*.

ARCH PAIN

Pain in the arch is almost invariably due to strain on the plantar fascia.

Plantar Fasciitis (*Very Common*)

Diagnosis

Pain and sometimes tenderness is present in the arch or the anterior portion of the bottom of the heel (Figure 20-2).

Pathophysiology

Strain of these ligaments is almost invariably due to excessive pronation. (Occasionally it may be seen in a high-arched foot.)

Treatment

Foot *hyperpronation must be controlled.* See p. 295.
Make sure that the sole of the shoe isn't too stiff.
Uphill running must be avoided.
Mileage should be decreased to where the patient is pain-free and then slowly increased, using alternate hard and light days.
Ice before and after running, a course of oral antiinflammatories or even a *steroid injection* (see Figure 18-4) may be used in more severe cases.

INFERIOR HEEL PAIN

Evaluation

See Figure 20-2.
If tenderness is actually at the plantar fascia or proximal arch, see PLANTAR FASCIITIS, above.
If tenderness is in the back or the bottom of the heel, see HEEL CONTUSION, immediately below.
If laterally compressing the calcaneus is painful, see CALCANEAL STRESS FRACTURE, below.

Heel Contusion (*Common*)

Diagnosis

Pain and tenderness are found in the posterior part of the inferior heel.

Pathophysiology

This is simply a result of direct soft tissue trauma from repetitive heel strikes.

Treatment

The shoe must have sufficient *padding* in the heel area; a *flared heel* is much more effective at dissipating shock. See Patient Handout 20-1.

Ice applied to the area before and after running will be helpful.

Mileage should be temporarily decreased and then reincreased as tolerated using an alternate day training schedule.

All the methods of decreasing transmitted force should be employed: running on *soft and level terrain and increasing pace.*

Calcaneal Stress Fracture (*Uncommon*)

Diagnosis

The patient will usually report that heel pain occurs with running, began after a dramatic increase in mileage, and is getting steadily worse.

Squeezing the calcaneus will be painful.

X ray may be revealing, but can be normal early in the course. A bone scan will usually demonstrate the pathology.

Pathophysiology

A result of repetitive compressive force.

Treatment

Running should be *stopped* for four to six weeks, and then resumed gradually, *as tolerated.*

All the measures to decrease transmitted force must be employed: *shoes that absorb shock well,* a brisk pace, and a *soft flat surface.*

ACHILLES AREA PAIN

Evaluation

See Figure 20-3.

A small tender bump over the achilles tendon is usually due to direct pressure

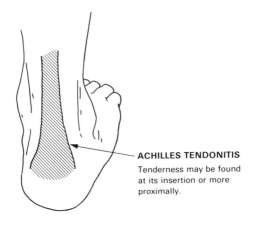

ACHILLES TENDONITIS
Tenderness may be found
at its insertion or more
proximally.

POSTERIOR VIEW

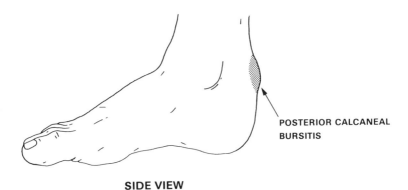

**POSTERIOR CALCANEAL
BURSITIS**

SIDE VIEW

Fig. 20-3.   Pain in the posterior heel.

from the counter of the shoe (POSTERIOR CALCANEAL BURSITIS). Padding in the area must be increased, or a new shoe that will not rub the area must be obtained. See also Chapter 18.

If tenderness is at the achilles tendon insertion, or more proximal and not over an inflamed bursa, see below.

Achilles Tendonitis (*Very Common*)

Diagnosis

Simply made by the presence of pain and tenderness along the achilles tendon or at its insertion.

Pathophysiology

Strain at this point can be due to several factors, singly or in combination: too low a heel on the shoe, excessive dorsiflexion due to uphill running, tight calf muscles, or a shoe with too little flexibility in the sole. Whether excessive pronation also can lead to problems here is a matter of debate.

Treatment

A *heel lift* should be inserted into both shoes; usually one-fourth of an inch will be about right.

Be sure that the *sole* of the shoe is *not too stiff.*

Running should be done only on *flat surfaces.*

Stretching exercises for the calf muscles must be done scrupulously. Once the symptoms have resolved, stretching for the achilles tendon should be added to help prevent recurrence (Patient Handout 20-2).

It is possible that *correcting hyperpronation* (as discussed on p. 295) will help.

Mileage should be *decreased* temporarily and then reincreased as tolerated, using an alternate-day training schedule.

In more severe cases, *ice* to the area before and after running and/or a course of oral *antiinflammatory medication* should be used. Steroid injection is contraindicated because of possible predisposition to rupture.

ANKLE SPRAINS (ACUTE AND RECURRENT)

These are discussed beginning on p. 217. Note that medial heel wedges, which are widely used and quite effective in the treatment of excessive pronation, can predispose to inversion sprains.

ANKLE PAIN

Evaluation

Bony tenderness must always bring to mind the possibility of a *stress fracture.* If X rays are normal and the patient resists the suggestion to discontinue running for six weeks on the basis of clinical evidence, a bone scan should be performed.

Tendonitis is much more common in this area. See Figure 20-4 and then the discussion of the appropriate condition below.

Diffuse ankle pain and/or tenderness, or pain on all directions of motion, may mean an intraarticular process. See Chapter 17.

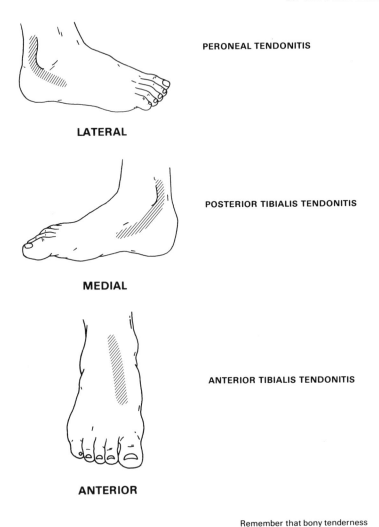

PERONEAL TENDONITIS

LATERAL

POSTERIOR TIBIALIS TENDONITIS

MEDIAL

ANTERIOR TIBIALIS TENDONITIS

ANTERIOR

Remember that bony tenderness
may mean a **STRESS
FRACTURE**. Also see Fig. 17-2.

Fig. 20-4.   Pain around the ankle.

Posterior Tibialis Tendonitis (*Very Common*)

Diagnosis

Tenderness over the posterior tibialis tendon is found (Figure 20-4), without
any bony tenderness.

Pathophysiology

This is due to foot hyperpronation pulling on the tendon. The posterior tibialis muscle contracts to try to resist the excessive pronation.

Treatment

Excessive pronation should be *corrected* as discussed on p. 295.

Check to be sure that there *is no pressure over the sore area* from the shoe.

Mileage should be temporarily *decreased,* and reincreased as tolerated using an alternate-day schedule.

In more severe cases, *ice* before and after running and/or a course of oral *antiinflammatory medication* can be used. *Steroid injection* around the tendon may be offered: see Chapter 3.

If the tendon attaches to a painful accessory navicular, *surgery* may be necessary.

Anterior Tibialis Tendonitis (*Common*)

Diagnosis

Tenderness is located over the anterior tibialis tendon (Figure 20-4), without any bony tenderness.

Pathophysiology

The anterior tibialis muscle must be contracted at the moment of heel strike to prevent the foot from immediately slapping on the ground; if it has to maintain its contraction through too long an arc because of excessive plantar flexion at the ankle (from a too-high heel or from running downhill), strain can occur. A habit of landing on the toes can also overstrain this muscle.

Treatment

A shoe with a *lower heel* should be selected and the patient advised to run only on *flat terrain.*

Mileage should be *decreased* to an asymptomatic level and then slowly brought back up using an alternate day training schedule.

Check to see that the tongue of the running shoe is not causing pressure on the painful area.

Stretching and *strengthening exercises* should be performed as in Patient Handouts 20-2 and 3.

In more severe cases, advise *ice* before and after running or oral *antiinflammatory medications*. *Steroid injection* around the tendon may be offered; see Chapter 3.

Peroneal Tendonitis (*Uncommon*)

Diagnosis

Tenderness is found over the course of the peroneal tendons (Figure 20-4).

Pathophysiology

Excessive *supination* leads to strain here, since the peroneal muscles will resist this motion. This is almost invariably found in the high-arch foot, but can be due to a running style of prolonged weight-bearing on the outside of the foot.

Treatment

Be sure the patient *is not running* on the *outside of his foot*.
A *lateral heel wedge* will help.
Mileage should be *cut back* to the point where symptoms abate, and slowly reincreased as tolerated using alternate-day training.

In more severe cases, *ice* packs before and after running and/or *antiinflammatory medications* can be used. *Steroid injection* around the tendon can be offered. See Chapter 3.

SHIN PAIN (SHINSPLINTS)

Evaluation

See Figure 20-5. Always check distal neurovascular status and consider the possibility of COMPARTMENT SYNDROME.

Diffuse tenderness over the medial distal tibia and adjacent muscle is found in POSTERIOR TIBIALIS STRAIN/PERIOSTITIS; marked or well localized tenderness may well signify a TIBIAL STRESS FRACTURE.

Pain over the antero-lateral leg implies ANTERIOR TIBIALIS STRAIN, which often recurrently or precipitously develops into a COMPARTMENT SYNDROME.

Tenderness localized over bone should be considered to indicate a STRESS INJURY (or when more severe a STRESS FRACTURE) OF THE TIBIA OR FIBULA until proven otherwise by bone scan.

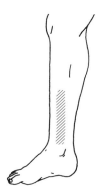

This is the usual location of pain and tenderness with inflammation of the POSTERIOR TIBIALIS muscle or its origin in the periosteum of the tibia. If it is *not clear* that the tenderness is *not* on bone, it may also mean a TIBIAL STRESS FRACTURE.

MEDIAL

This is the usual area in which STRESS FRACTURES OF THE FIBULA occur.

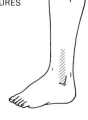

LATERAL

Tenderness on bone can mean a STRESS FRACTURE OF THE TIBIA.

ANTERIOR

Tenderness here means strain of the ANTERIOR TIBIALIS muscle, which can intermittently or precipitously progress into an ANTERIOR COMPARTMENT SYNDROME.

Tenderness here means strain of the PERONEAL MUSCLES, or if not clearly distinguishable from bone could mean a STRESS FRACTURE OF THE FIBULA.

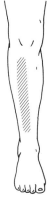

ANTERIOR

LATERAL

Fig. 20-5. Pain in the shin.

Tenderness posterior to the distal fibula is found in strain of the peroneal muscles; see the discussion of PERONEAL TENDONITIS earlier in the chapter.

Posterior Tibialis Strain/Periostitis (*Very Common*)

Diagnosis

Pain and tenderness are found over the medial lower leg (Figure 20-5). If there are any focal areas of bony tenderness or if there is significant diffuse bony tenderness, a STRESS FRACTURE OF THE TIBIA may be present; X ray and, if that is normal, a bone scan must be performed to exclude this more serious diagnosis.

Pathophysiology

The posterior tibialis resists excessive foot pronation. Thus, strain of this muscle or inflammation of its periosteal origin on the tibia is a result of hyper-pronation.

When severe, such strain on the muscle's bony origin can result in stress injury (or eventually, stress fracture) of the tibia itself.

Treatment

Hyperpronation must be *controlled* as discussed in the opening of the chapter. *Mileage* should be *decreased,* and then slowly re-increased as tolerated. Stress fractures should be managed as discussed below.

Anterior Tibialis Strain (*Common*)

Diagnosis

Pain and sometimes tenderness will be found over the anterolateral leg (Figure 20-5).

Because of the tightness of the anterior compartment of the leg, overuse of this muscle often leads to intermittent mild ANTERIOR COMPARTMENT SYN-DROMES. A persistent or progressive compartment syndrome is of course an indication for emergent orthopedic consultation.

Be careful to exclude bony tenderness which would indicate STRESS INJURY (OR FRACTURE) OF THE TIBIA.

Pathophysiology

This muscle must undergo a lengthening contraction from the moment of heel strike until the forefoot is solidly on the ground, to prevent an uncontrollable "slapping" gait. Going downhill or an excessively high heel thus result in a lengthened arc through which the anterior tibialis must maintain its contraction.

Trying to avoid slipping while running on wet or icy surfaces also often results in overuse of this muscle.

Some athletes may have an especially tight anterior compartment and have much trouble with intermittent compartment syndromes.

Treatment

Lower the heel a bit and advise the patient to *avoid running downhill.*
Stretching exercises, and *ice* after running, may be helpful.

Serious athletes who have much trouble with recurrent anterior compartment swelling in spite of these measures may be candidates for elective fasciotomy.

Stress Fracture of the Tibia or Fibula (*Common*)

Diagnosis

The patient will give a history of pain worse after running and relieved by rest, but of course this is not characteristic of only this problem.

On examination, bony tenderness will be found. Often it will be hard to distinguish tenderness of the bone from that of adjacent or overlying muscle, in which case work-up must be done to exclude the more serious diagnosis.

The X ray may be characteristic, or it can be normal early.

A bone scan will light up in the involved area(s).*

Pathophysiology

See the discussion in Chapter 1.

Tibial stress fractures are more common than those of the fibula.

Stress fractures of the tibia are probably usually due to transmitted axial stress,

---

*Radionuclide uptake in a longitudinal pattern along the tibia can be consistent with a periostitis at the origin of the posterior or anterior tibialis muscles rather than a stress fracture. The latter will usually produce one or more very focal or transverse areas of uptake. It is conceivable however, that the former may predispose to, or progress to, the latter.

but it is quite possible that stress on the origins of the posterior tibialis (posterior proximal tibia) or anterior tibialis (anterolateral proximal tibia) muscles may be contributing factors at least in some cases.

Stress fractures of the fibula are harder to attribute to transmitted weight, since this bone absorbs only about 15% of such stress. So, most likely, stress at the origins of the peroneal muscles (or in the case of the fibular head, at the insertion of the biceps femoris) is responsible.

Treatment

The patient with a stress fracture must *discontinue running* (and prolonged walking on hills or hard surfaces) for *at least six weeks,* and must be informed that if he or she does not follow this prescription, the stress fracture can at any time suddenly turn into a real through and through break, with all the complications that that implies.

Crutches can be used for a few days or weeks if pain is severe; occasionally the patient will prefer the application of a cast, but this is a symptomatic measure only.

The patient can resume weight-bearing as the pain improves. If at any time during recovery an advance in the amount of weight-bearing leads to a recurrence of pain or tenderness, the patient should revert to the previous pain-free level of activity for a couple of weeks before trying to advance his or her weight-bearing again.

Cycling is often a well-tolerated intermediate step before resuming running.

After six weeks or so, if the patient has been pain-free for at least two weeks, running can be resumed, but very, very slowly. The training regime should be cut back or interrupted at the first sign of pain. Shoes must have *good shock absorption.* and running must be done only on soft, flat terrain at a not-too-slow pace, using the alternate day regimen. If because of the location of the fracture or concomitant pain in adjacent musculature, stresses on muscular origins are thought to have played a part, the *specific biomechanical corrections* discussed in the appropriate section should be applied as well.

Stress injury is a term used to denote milder degrees of bony damage. The scan will not be as "hot" as with a fracture. The pathophysiology is of course identical with that of stress fracture (into which it develops if excessive stress is not relieved.) Treatment principles are also identical, except that, depending on severity and progress, abstention from activity *may* not have to be as complete or prolonged.

Compartment Syndromes (*Rare*)

Diagnosis

Although seen perhaps more commonly after severe trauma, vascular occlusion or orthopedic surgery, compartment syndromes of the lower leg can also occur as a result of exercise.

The patient will complain of pain in the involved compartment and sometimes will complain of a neurologic deficit as well. By far the most common exercise-induced compartment syndrome involves the anterior compartment, which can give numbness in the first web space and weakness of the toe extensors. (*Note:* A decrease in two-point discrimination is the earliest, most subtle sign of neurologic compromise.)

Tenderness and sometimes palpable tenseness will be found over the affected compartment, and pain will result from *passive* stretch of the muscles there. (In the case of the anterior compartment, passive toe and foot plantar flexion will hurt.)

This problem can present in an acute and progressive way (in which case the patient must be seen by an orthopedist *STAT*), or, more commonly when running-induced, as a mild, self-resolving, recurrent complaint after exercise, often with no findings at all when seen by the doctor.

Pathophysiology

In the cases under consideration here, increased capillary permeability due to exercise leads to swelling and thus increased pressure in the closed space. This can progress to ischemia and thus set up a vicious cycle.

The increased pressure then can lead to peripheral nerve compression and destruction if it is not resolved spontaneously or surgically.

Treatment

Acute or progressive cases or those in which neurologic findings are present at the time of examination *must* be evaluated by an orthopedist within no more than a few hours.

Patients who give a history suggestive of mild recurrent compartment syndrome should be aggressively treated to remove excessive strain on the muscle groups involved, as discussed in the appropriate section.

CALF PAIN

Evaluation

Sudden onset of calf pain, sometimes with a felt or heard snap and sometimes with mild associated swelling, is indicative of a PLANTARIS RUPTURE. See Chapter 16.

Always keep in mind the possibility of a *deep vein thrombosis* and, in older patients, of *claudication* due to vascular disease. The diagnosis of these conditions is beyond the scope of this book, and indeed is often difficult, but their possibility must not be left unconsidered.

Also consider the possibility of calf pain being *referred* from the *low back*.

Once you are fairly certain that these conditions are not the problem, see the discussion that follows.

Calf Strain (*Common*)

Diagnosis

Pain and tenderness may be located anywhere in the calf, from the origin of the muscles in the back of the knee to where they become the achilles tendon.

Be sure that the clinical picture is not compatable with DVT, claudication, plantaris rupture, or referred pain.

Pathophysiology and Treatment

See the discussion of ACHILLES TENDONITIS, above.

Stretching exercises for the calf muscles (and *strengthening exercises* as well) are especially important. See Patient Handouts 20–2 and 3.

Acute cases should be treated with *cold* packs, a period of *rest*, and, possibly, *antiinflammatory medication*.

KNEE PAIN

Evaluation

Knee sprains and other trauma, knee swelling, and the problem of the collapsing knee are discussed in Chapter 15.

Knee pain in runners is usually caused by the conditions discussed in this section, but can certainly be due to other conditions (especially meniscal tears) as well. If one of the syndromes discussed here does not seem to fit quite right, refer to Chapter 15 for a discussion of a more complete knee evaluation.

Keep in mind the possibility of pain referred from the hip or low back.

If there is bony tenderness, the patient may very well have a *stress fracture* (most commonly in the proximal tibia). This is a diagnosis that must not be missed: see further discussion earlier in this chapter.

See Figure 20-6 for the location of pain and tenderness with various other conditions and then the discussion of the appropriate syndrome.

Patellar Tendonitis (*Common*)

Diagnosis

Pain and sometimes tenderness will be found at either end of the patellar tendon (Figure 20-6). Be careful that the patient does not have bony tenderness that could mean a proximal tibial stress fracture.

Pathophysiology

At the moment of heel strike, the quadriceps must suddenly and forcibly contract to prevent the knee (which is flexed at this juncture) from collapsing.

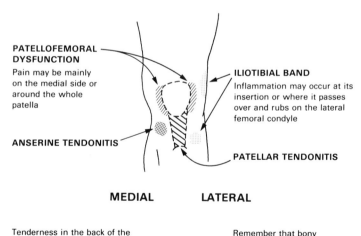

Fig. 20-6. Pain around the knee.

This contraction is transmitted to the tibia through the patellar tendon, and so inflammation at this site can develop. Thus, this is a problem of excessive transmitted weight.

Treatment

All of the measures to decrease transmitted stress should be applied: *shoes with better shock absorption, avoidance* of *hills* and of *hard surfaces,* and increased speed.

Mileage should be decreased and then slowly brought back up, using an alternate-day training schedule.

Quadriceps stretching exercises (Patient Handout 20-2) are important for prevention of recurrence.

Ice to the area before and after running and/or a course of *antiinflammatory medication* can be used in more severe or acute cases.

Steroid injection is contraindicated because of the great stresses at this point and the possibility of subsequent rupture.

Patellofemoral Dysfunction (Chondromalacia) (*Very Common*)

Diagnosis

The patient will usually complain of a vague ache in the front of the knee, and sometimes of a grating sensation as well. Occasional episodes of collapse may punctuate this syndrome, but if they are present a more detailed examination to rule out other potential pathology should be made. (See Chapter 15.)

On examination, there will be pain and sometimes palpable crepitance when the patella is pushed down against the femur and moved up and down.

Peripatellar tenderness may be present, usually mainly on the medial side.

If the problem is mainly on the medial side, the patellar compression test may be more positive if the patella is pushed medially as it is compressed, and passive foot pronation may result in knee pain.

Pathophysiology

The patella acts as a fulcrum for the extensor mechanism of the knee, so when at the moment of heel strike the quadriceps forcefully contract to prevent giving way of the then-flexed knee, the kneecap is squashed into the femur. This can result in pain and eventually in roughening of the cartilage as well. (Previously existing chondromalacia patellae or patellar tracking problems due to weak quadriceps or congenital factors can certainly predispose to the development of this problem in the runner.)

A completely different mechanism is operative in some cases: excessive foot *pronation* leads to excessive internal torsion of the leg, which compresses the medial side of the patella against the medial femoral condyle. Of course, both mechanisms may coexist and be contributory in the same patient.

Treatment

Especially if the medial part of the patellofemoral articulation is primarily involved, foot *hyperpronation* must be *corrected* as discussed on p. 295.

Measures to decrease weight transmission should be employed: *shoes which absorb shock well,* limitation to *flat terrain* with *soft surfaces,* and a brisk pace.

Quadriceps strengthening exercises as shown in Patient Handout 15-2 may be helpful to improve patellar tracking.

Distance should be temporarily *decreased* and then slowly increased, alternating hard days with light days.

A course of *antiinflammatory* medication can help in more severe cases.

Iliotibial Band Tendonitis (*Very Common*)

Diagnosis

Pain and sometimes tenderness will be found where the iliotibial band inserts or more commonly where it passes and rubs over the lateral femoral condyle (Figure 20-6). At the latter site, crepitance can often be felt as the knee slightly flexes and extends while close to thirty degrees of flexion.

The same structure can be painful along its course in the lateral thigh, or where it passes over the greater trochanter, or at its origin (as the tensor fascia lata muscle) at the anterior superior iliac spine.

Having the patient lie on the unaffected side, extending the higher hip and then forcing it downward (thus scissoring the legs) will usually be painful (Ober's sign.)

Iliotibial band problems are a very common overuse syndrome in cyclists, for obvious reasons.

Pathophysiology

People run with their feet striking the ground well toward the midline. Therefore at the moment of heel strike there is a varus stress at the knee, and the iliotibial band is pulled by this, and it is also pulled by its tensor fascia lata muscle attempting to resist the motion.

Running on a slope will greatly increase strain on the IT band in the higher leg. If the feet strike too far toward the midline, or even cross slightly (called a scissoring gait), stress on this structure will be increased.

It is conceivable that hyperpronation and subsequent internal rotation of the leg increases traction on this structure.

Treatment

Running on a slope (i.e., with one foot higher than the other) must be stopped.

A *scissoring gait* must be corrected; even in runners who do not scissor, running with the feet striking the ground just a little further from the midline may help.

Measures to decrease transmitted weight should be applied (running only on *flat, soft terrain* with *well-cushioned shoes* at a fairly good pace).

As always, *distance* should be *cut back* and relengthened gradually using alternate heavy and light days.

Ice before and after running and/or oral *antiinflammatory medications* can be prescribed. Steroid injection at the point of maximal tenderness can be offered in recalcitrant cases (See Figure 20-7).

A trial of *correcting* any *hyperpronation* that may be present may be worthwhile.

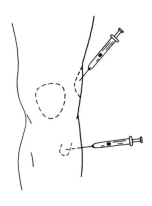

MEDIAL   **LATERAL**

Fig. 20-7. Injecting the iliotibial band. After prepping the area and spraying with ethyl chloride, use a 25-gauge needle to infiltrate a cc of steroid and a cc of 1% xylocaine around the tender area. See p. 23 for further details and precautions.

Anserine Tendonitis (*Common*)

Diagnosis

Pain and sometimes tenderness are present at the conjoined insertion of the sartorius, gracilis, and semitendinosus muscles (Figure 20-6).

Pathophysiology

Inflammation there results from excessive valgus strain at the knee, most commonly from running on an inclined surface (such as the side of a road) with one leg higher than the other.

Treatment

Running on *inclined surfaces* should be *stopped*.

Check for and *correct* with a lift about one-half of any *leg length discrepancy*.

As always, *distance* should be *cut back* and slowly reincreased as tolerated using alternate-day hard and light training.

A *cold pack* applied before and after running and/or oral *antiinflammatories* can be tried.

If these measures do not bring relief, the patient should try running with the foot striking the ground just a little closer to the midline.

Steroid injection has a good success rate. See Figure 15-12.

THIGH PAIN

Evaluation

See Figure 20-8.

Pain and tenderness over the greater trochanter is discussed under TROCHANTERIC BURSITIS, below.

Pain and/or tenderness diffusely along the lateral side of the thigh is probably in the iliotibial band; see the discussion of ILIOTIBIAL BAND TENDONITIS in the previous section.

Pain in the quadriceps muscles in the front of the thigh is discussed below, as are pain in the adductor muscles on the medial side and pain in the hamstring muscles in the posterior thigh.

Keep in mind the possibility of pain referred from the hip or low back, or of a (rare) STRESS FRACTURE OF THE FEMUR.

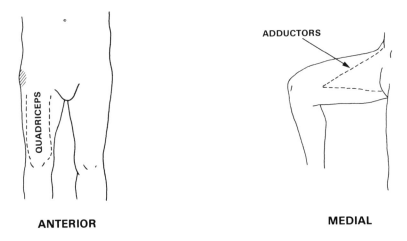

ANTERIOR   **MEDIAL**

If the diagnosis isn't obviously a muscle or iliotibial band problem, consider the possibility of a FEMORAL STRESS FRACTURE.

Indicates the GREATER TROCHANTER, into which muscles which balance the pelvis insert and over which the iliotibial band passes.

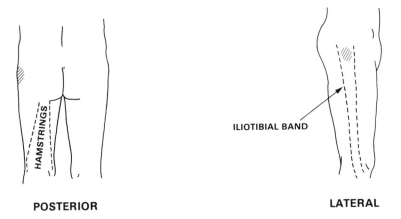

POSTERIOR   **LATERAL**

Fig. 20-8.   Pain in the thigh.

Trochanteric Bursitis (*Common*)

Diagnosis

Pain and tenderness are located over the greater trochanter of the femur (Figure 20-8).

Pathophysiology

Although commonly called a bursitis, three different structures can produce pain at this site. The *gluteus medius* muscle inserts here. Contraction of this muscle is necessary to balance the pelvis while the runner is in the stance phase on the other leg; this is especially likely to lead to insertional strain when running on an inclined surface with one leg higher than the other, or if there is a leg length discrepancy. The *iliotibial band* passes over the trochanter and can become inflamed from its back and forth motion over the bone. Or a *bursa* may form underneath it and become inflamed.

Treatment

Running on *inclined surfaces* should be *stopped*.

If a *leg length discrepancy* is found, 50% of the difference should be *corrected* with a lift.

A temporary *decrease* in *mileage* should be prescribed, with slow reincrease using alternate hard and light days.

If these do not solve the problem, the measures discussed under ILIOTIBIAL BAND TENDONITIS in the previous section should be tried.

Steroid injection may be offered. See Figure 14-4.

Quadriceps Strain (*Uncommon*)

Diagnosis

Pain is located in the muscles of the anterior thigh.

Pathophysiology and Treatment

See the discussion of PATELLAR TENDONITIS earlier in the chapter. *Stretching exercises* are especially important (Patient Handout 20-2), and strengthening exercises should be performed as well, since relative weakness of the quadriceps as compared to the hamstrings may be a contributing factor.

A slightly *longer stride* may help.

Adductor Strain (*Uncommon*)

Diagnosis

Pain is located in the medial thigh or the groin.

Pathophysiology

This is due to running with the feet striking too far from the midline, or more commonly to running on an inclined surface.

Treatment

Distance should be *cut down* and then built up slowly, using the alternate hard and light day schedule.

Running should be restricted to *level surfaces,* and *foot strike* encouraged to be *more toward* the *midline.*

Correct half of any *leg length discrepancy.*

Stretching exercises of this muscle group should be performed faithfully to prevent recurrence (Patient Handout 20-2).

Cold packs before and after running or a course of *antiinflammatory* medication can be used, especially in acute strains.

Hamstring Strain (*Common*)

Diagnosis

Pain is located anywhere in the posterior thigh, or at the origin of the hamstrings in the buttock, or at their insertions in the back of the knee. (Keep in mind the possibility of referred sciatic pain.)

Pathophysiology

At the moment of heel strike, the quadriceps must forcefully contract to prevent collapse of the partially flexed knee; the hamstrings then must immediately begin contracting to continue the stride. Therefore, quadriceps muscles too strong relative to their hamstrings, or a stride that is too long and forces the hamstrings to begin their contraction from a position of extreme length, can lead to problems.

Treatment

Distance should of course be *cut back* and slowly reincreased using an alternate day regime.

Stretching and strengthening exercises as shown in Patient Handouts 20-2 and 20-3 are very important in rehabilitation and prevention of recurrence. Aim for a quadriceps: hamstrings strength ratio of 3:2.

A slightly shorter stride may help.

In acute strains, *cold packs* and *antiinflammatories* are especially useful; they may be of benefit in chronic strains as well.

GROIN PAIN

Evaluation

Groin pain may be a manifestation of STRESS FRACTURE OF THE FEMORAL NECK. X ray and bone scan must be obtained to rule out this possibility, since the results of attempting to run through the pain could be *catastrophic*. If found, the condition should be managed by the orthopedist.

More commonly, groin pain will be due to strain at the origin of the adductor muscles; this sometimes may be so severe as to cause a periostitis or actual stress fracture. See ADDUCTOR STRAIN, discussed in the previous section.

BUTTOCK PAIN

This is usually a hamstring strain, but consider and rule out the possibility of pain referred from the back or the hip.

PUBIS PAIN

Excessive bilateral abduction strain can predispose to OSTEITIS PUBIS, visible on bone scan or eventually on X ray. (This can occur in long-distance runners simply from shearing stress at the pubic symphysis as the pelvis moves up and down.) For treatment, see the discussion of ADDUCTOR STRAIN on p. 320.

ILIAC PAIN

This is the origin of the tensor fascia lata, which becomes the iliotibial band; see the discussion of strain of that structure, above. Note that in teenage runners an *avulsion fracture* may occur here, in which case discontinuance of running for about six weeks is mandatory, followed by slow resumption while paying heed to the other factors discussed in the section on ILIOTIBIAL BAND TENDONITIS.

UNILATERAL NECK PAIN

This is often due to running on an inclined surface, with one foot hight than the other and the neck tilted to compensate.

Patient Handout 20-1

What to Look for in a Running Shoe

Note: Good shock absorption capacity is particularly important if your feet are rigid. If you have a flexible foot, controlling hyperpronation by holding the heel snugly in place and supporting the arch is essential. These are general guidelines; specific adjustments may have to be made for your anatomy.

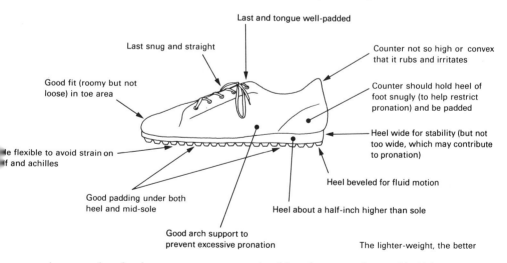

Last and tongue well-padded

Last snug and straight

Counter not so high or convex that it rubs and irritates

Good fit (roomy but not loose) in toe area

Counter should hold heel of foot snugly (to help restrict pronation) and be padded

Heel wide for stability (but not too wide, which may contribute to pronation)

Sole flexible to avoid strain on calf and achilles

Heel beveled for fluid motion

Good padding under both heel and mid-sole

Heel about a half-inch higher than sole

Good arch support to prevent excessive pronation

The lighter-weight, the better

As your shoe begins to wear out, you should make corrections with "shoe-glue." But *never overcorrect.*

Patient Handout 20-2

Stretching Exercises for Runners

Warm up for a few minutes by going much slower than your usual pace or by running in place. Then do each exercise by slowly stretching (not bouncing) to the point of tightness (not pain). Hold the stretch for 30 to 60 seconds.

When you have finished running, a gradual cool down and repeating your stretching exercises will help prevent soreness and injury.

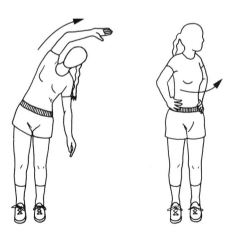

SIDE BENDS   TRUNK TWISTS

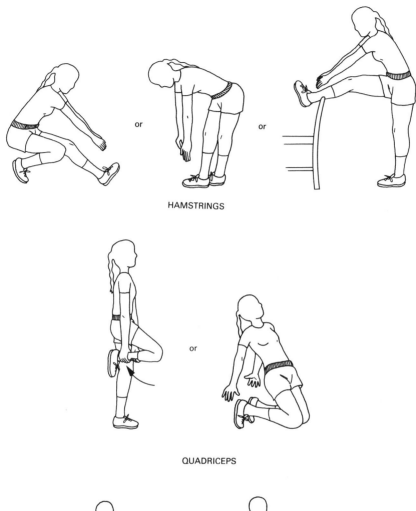

HAMSTRINGS

QUADRICEPS

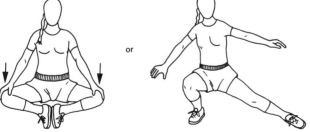

GROIN AND ADDUCTORS

CALF

ACHILLES

Patient Handout 20-3

Strengthening Exercises for Runners

Because running involves the muscles in the back of the legs so much, you can develop a relative weakness of the muscles of your abdomen and the front of your thigh. So it's not a bad idea to perform the following exercises to keep them strong.

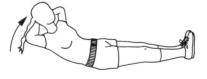

ABDOMINALS

Do 15 of each every day.

Do them slowly; as you get stronger you may want to progressively add more weights to your ankles.

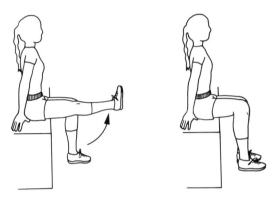

QUADRICEPS

If your knee hurts, do this
in a different way:

You need to do the following ones only if your doctor recommends. (Do 15 of each every day.)

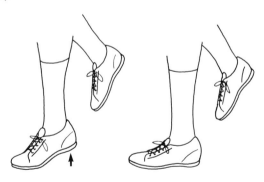

☐ CALVES

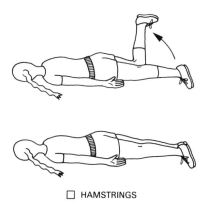

☐ HAMSTRINGS

Do this slowly; as you get
stronger you may want to
progressively add weights
to your ankles.

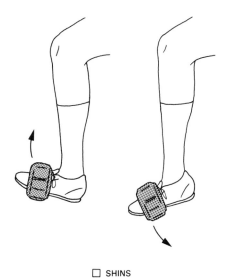

☐ SHINS

Patient Handout 20-4

A Running Prescription

(Check all that apply.)

☐ Decrease mileage to a point where symptoms disappear and then slowly bring it back up as tolerated. Alternate heavy training days with light ones or with days off. Don't do "interval training."

☐ Be sure to do stretching exercises after proper warm up, and again after running. Cool down gradually.

☐ Pay particular attention to the exercises checked.

☐ Put an ice pack on the sore area for ten minutes before and after running.

☐ Take medication as directed.

^{w}☐ Don't run on hilly terrain.

^{w}☐ Don't run on hard surfaces.

^{w}☐ Increase your speed (without affecting your mileage) as much as is practical.

^{p}☐ Don't run on inclined surfaces (with one leg higher than the other) or around a small track.

☐ Adjust or change your shoes as follows:

    ☐ raise heel   ☐ lower heel

    ^{p}☐ snug heel and better arch support ^{p}☐ obtain customized orthotic

    ^{w}☐ better shock absorption ☐ _____

---

*Note: "w" indicates measures to decrease transmitted weight; "p" indicates measures to decrease excessive pronation.

21

Muscle Problems

The differential diagnosis of *diffuse muscular weakness* obviously includes various diseases of muscle, the motor end plate, and the peripheral and central nervous system, and is beyond the scope of this chapter. The reader is referred to textbooks of internal medicine or neurology for discussion of the possibilities. Keep in mind that unexplained *muscle masses* can be *malignant.*

MUSCLE TRAUMA

Muscle Pull (Strain) (*Mild Ones Extremely Common, Severe Ones Uncommon*)

Diagnosis

   The history is of a single excessive forceful stretch or of repeated mild excessive stretch of the involved muscle.

There will be local pain and tenderness in the involved area, and the involved muscle may be in spasm.

Ecchymosis and/or impaired strength of the involved muscle indicate at least some tearing of the muscle (this is called a *second degree strain*). If there is a palpable defect in the muscle, or if it is noted to "bunch up" toward either its origin or insertion, a complete tear (*third degree strain*) is present, and there will be significant loss of function.

Check distal neurovascular status and keep in mind the possibility of a compartment syndrome.

Pathophysiology

Acute strains are usually the result of a sudden stretching force for which the muscle is not prepared, or which is greater than its ability to resist.

Chronic strains are a result of the muscle having insufficient flexibility for the stretch being repetitively applied. An imbalance of strength with its counteracting muscle can also be a contributing factor.

As noted above, first degree strains consist of stretched muscle, second degree strains of partial tears, and third degree strains of complete tears.

Treatment

An acute first degree or mild second degree strain should be treated with *ice, compression* and *elevation,* and it should be *rested* until symptoms diminish.

Following such an acute strain, a program of nonballistic *stretching exercises* should be carried out to help prevent recurrence.

In chronic strains, exercises to stretch the involved muscle are most important. *Strengthening* the muscle in relation to its counteracting muscle is often helpful as well.

(For example, repetitively strained hamstrings may be too weak in comparison to the quadriceps.)

Severe second degree and all third degree strains should be referred to an orthopedist, since often *surgical repair* is warranted, and the decision whether or not to operate is often a complex one. If repair is to be done, it must be performed immediately.

Contusion (Bruise) (*Extremely Common*)

Diagnosis

After direct trauma, ecchymosis develops. If bleeding is deep into a large muscle mass, though, discoloration may not be evident, at least early. In this case the history and the complaint of deep pain increased with motion, along with a negative X ray, is sufficient to make the diagnosis.

If bruising is repeated, excessively easy or unexplained, the possibility of a clotting problem should obviously be considered and investigated. In children, keep in mind the possibility of child abuse.

Pathophysiology

The hematoma can be subcutaneous or within the body of the muscle. Less commonly it is subperiostial.

The hematoma eventually is "organized" by the development of granulation tissue and then fibrous scar tissue.

Treatment

Acutely, *ice, compression,* and *elevation* should be used.

Very large bruises that are not clearly superficial should be *rested for several weeks* to help prevent the development of MYOSITIS OSSIFICANS. *Massive* hematomas may need to be *surgically drained.*

Myositis Ossificans* (*Rare*)

Diagnosis

Pain and tenderness persist in the area of a large hematoma, most commonly in the thigh or upper arm.

Calcification can be evident on X ray a few weeks after injury. Ossification is usually complete in several months. If large, the bony mass is palpable.

This condition can be confused with bone tumors; obviously referral for biopsy is necessary if there is any doubt.

Pathophysiology

An organizing hematoma becomes calcified and is then invaded by osteoblasts which form new bone. The precise trigger and mechanism are unknown.

---

*The condition discussed here is classified as *myositis ossificans circumscripta* or *myositis ossificans traumatica,* because it consists of the ossification of an organizing hematoma. There is a quite distinct disease of unknown etiology called *myositis ossificans progressiva* or *generalized myositis ossificans,* which usually strikes children or young adults, is often associated with congenital deformities, and progresses in relentless fashion to involve the entire body. No effective treatment is known.

Treatment

The involved muscle should be rested (*not* immobilized) for several months, and then gentle strengthening exercises begun.

Otherwise only *symptomatic measures* are necessary unless significant disability persists, as is often the case if the mass is near a joint. Then the mass can be surgically removed once maturation is complete (generally about a year post-injury).

DIFFUSE MUSCLE PAIN

Evaluation

Obviously, by far the most common cause of acute diffuse myalgia is acute systemic infection, usually viral. Keep in mind that acute diffuse muscle inflammation, whether from viral infection or overuse, can sometimes be severe and can lead to myoglobinuria and acute renal failure.

Long-term chronic diffuse muscle aching is usually related to deconditioning, degenerative arthritis with secondary muscular effects, or to psychological factors.

Subacute diffuse myalgia can be the presenting symptom of rheumatoid arthritis, other rheumatic diseases, or polymyositis. Other evidence for these conditions should be sought. But consideration must always be given to the possibility of POLYMYALGIA RHEUMATICA, discussed below.

Polymyalgia Rheumatica (*Rare*)

Clinical Features and Diagnosis

Usually occurs in patients above the age of 50.

The patient complains of a diffuse aching pain, most commonly in the neck and shoulders, and sometimes in the hips, thighs and arms as well. The pain is worse after motion, and there is morning stiffness.

The sedimentation rate is significantly elevated. The RF and ANA tests are negative.

A large number of these patients have associated temporal arteritis. This is of course potentially a much more serious condition. Opinion differs as to whether routinely to perform a temporal artery biopsy on all patients with PMR before instituting treatment, as arteritis can be spotty, leading to a false-negative result.

Pathophysiology

Unknown.

Treatment

Oral steroids are used. Because of the occasional difficulty in excluding temporal arteritis, and because the latter condition requires much higher dosages of these dangerous medications, *referral* to a rheumatologist is recommended.

MUSCLE CRAMPS (*Very Common*)

Diagnosis

The patient reports sudden contractions of a muscle with the development of a muscular knot, lasting up to several minutes.

The calf and the foot are the most commonly involved areas, but the cramps can occur anywhere.

They are most commonly due to activity excessive for the condition of the muscle, and occur either during the activity or hours later.

They also occur frequently in pregnant women secondary to lack of sufficient calcium intake.

Pulses and capillary filling must be checked to rule out vascular compromise as the cause of the muscular irritability, especially in older patients.

Tests for serum sodium, potassium, calcium, magnesium, phosphorus, and sugar and a CBC should be performed in patients with frequent recurrence, since diabetes, polycythemia, or abnormalities of electrolytes can be responsible.

Treatment

The patient should be instructed that when a cramp develops he should keep the muscle *stretched* until it abates. (In the case of the calf, by dorsiflexing the foot with the hand or by pushing against something with the forefoot).

Specific etiologies discovered in the evaluation discussed above should be dealt with.

If none of these are found, the patient should perform *stretching exercises* each morning. Teach or check the calf stretching exercises in Patient Handout 20-2.

Quinine sulfate (5 grains at bedtime) has been used with some success in the prevention of nocturnal cramps.

22

Degenerative Arthritis and Osteoarthritis

These terms are used interchangeably to describe a single pathologic entity, but they can be considered to represent two quite distinct (though often coexistent) clinical syndromes.

CLINICAL FEATURES

Degenerative Arthritis

This is the wearing-out of one or more joints. It is a ubiquitous and (so far) unavoidable consequence of aging.

Joints which have been overused or placed under more strain wear out sooner.

Joints which have been directly traumatized or had their mechanics altered by trauma, deformity or previous disease (such as rheumatoid arthritis or aseptic necrosis) likewise will degenerate before other joints in that individual.

The hip, knee, first MCP and first MTP joints, and the facet joints of the spine are most commonly involved, but any and all joints can become symptomatic.

Patients complain of aching pain in the joints, worse after usage or weight bearing; there is some stiffness after rest but this is not nearly as dramatic as in patients with rheumatoid arthritis.

A severe progressive form occurs in joints which lack sensory innervation because of neurologic disease. These are called *Charcot Joints*. (They are not usually painful.)

Osteoarthritis

This condition mainly affects small joints, most commonly the DIP's, where bony hypertrophy at the sides of the joints produces Heberden's nodes.

Less commonly, the PIP's are affected, giving rise to Bouchard's nodes.

This is primarily a disease of women; it can begin as early as the fourth decade.

There is a definite familial predisposition.

Pain and stiffness are usually much milder than in rheumatoid arthritis.

An uncommon form called *erosive osteoarthritis* can occur, with a single small joint or several becoming acutely inflamed and painful for several weeks.

DIAGNOSIS *

There are no systemic manifestations.

On examination, there may be joint line tenderness, pain on (and sometimes limitation of) range of motion, and occasionally crepitus. In joints in which the condition is more advanced, the bony deformity will be obvious.

Redness and warmth are lacking (except in erosive osteoarthritis).

X ray may be normal early, when the disease is confined to the articular cartilage; but later it becomes characteristic, showing joint space narrowing, bony hypertrophy (especially at the joint margins), and sometimes periarticular sclerosis without osteoporosis. A word of caution: these radiographic changes are ubiquitous in older patients and previously injured joints, and their presence does not rule out another cause for the specific complaint under investigation.

Blood tests, including the ESR, are normal.

Synovial fluid analysis reveals findings typical of noninflammatory conditions, with white cell counts in the 500 to 5000 range and consisting mostly of mononuclear cells.

*The material from this point through page 339 applies to both the degenerative and osteoarthritic forms of the disease.

PATHOPHYSIOLOGY

The first change found is in the articular cartilage, which becomes yellowed and focally softened and roughened. These changes then spread to involve the entire joint surface, which becomes progressively more frayed and cracked as time goes on.

Along with these changes, there occurs remodeling of the periarticular bone, with sclerosis of the bone underlying the articular cartilage and development of spurs of bone (osteophytes) at the joint margins as well as at the sites of attachment of tendons and ligaments. These changes may be a result of the normal remodeling process that occurs in bone with changes in applied stress, in this case the changes in applied stress being secondary to alterations in the configuration of the joint.

Synovial inflammation, the level of which is quite variable, can result from the stresses of altered joint mechanics, or from direct irritation of the synovial membrane by tiny fragments of articular cartilage released into the joint space.

Eventually all the articular cartilage can be worn away, leaving a joint consisting of bone on bone. Large fragments of cartilage and bone can become loose bodies within the joint, leading to locking and collapsing.

Various progressive histologic and biochemical changes have been observed in involved cartilage, but the precise mechanism that leads to the gross pathologic changes that are seen remains a mystery.

A break in the surface of the articular cartilage (due to microtrauma or macrotrauma) may allow proteolytic enzymes in the synovial fluid to enter and begin to digest the mucopolysaccharides, the beginning of a vicious cycle.

There is definitely a change in the function of the chondrocytes at the base of the cartilage; whether this is an attempt to repair the damage or leads to progression of it (or both) is still unknown.

Since cartilage is avascular and must receive its nutrition from the synovial fluid with which it is in contact, interruption of this mechanism can lead to degeneration. Perhaps this is why prolonged immobilization of a joint, which eliminates the "pumping" action provided by the intermittent compression that occurs in its normal use, leads to accelerated degenerative change.

A large part of the pain may actually arise from fatigue and tension in muscles surrounding the joint, rather than from joint inflammation itself.

TREATMENT

It is often important to *reassure* the patient that he does not have a progressive systemic disease, lest the diagnosis of arthritis lead to visions of wheelchairs and such (Patient Handout 22-1).

A proper *balance of rest and use* of the joint is important for symptomatic relief, and probably to help slow progression of the disease as well. The specific amount of each is best determined simply by the patient's symptoms. Too much use or not enough will lead to worsening of discomfort and stiffness. It must be emphasized to the patient that if his joints are hurting, they are telling him something; they need to rest.

Weight loss is very important when obese patients have involvement of the weight bearing joints.

Local *heat* has an analgesic effect (see the discussion in Chapter 3).

Salicylates and other *analgesic/antiinflammatory* medications are mainstays of therapy. Again, see Chapter 3 for more details.

Specific measures to avoid stress to involved joints (such as use of a cane in degenerative arthritis of the hip or a cervical collar when the neck is symptomatic) are discussed in the sections on degenerative arthritis in the chapters on the individual joints.

Steroid injection into selected troublesome joints can give temporary but prolonged relief. However, it is quite possible that this is at a cost of accelerating cartilagenous destruction and thus the underlying disease. An individual joint should probably not be injected more than two or three times a year. The procedure is discussed further in Chapter 3, and techniques for entering specific joints will be found in the corresponding chapters.

In moderately severe cases, referral to a *physical therapist* is advisable for training in range of motion and strengthening *exercises* to help maintain joint mobility and stability, and to restore proper joint mechanics as much as possible. The therapist can also be helpful in the provision of warming modalities such as ultrasound and paraffin baths.

Referral for *surgery* should be considered in severe disease which does not respond to the conservative measures described above. Several types are performed, depending on the individual case: joint debridement, joint fusion, osteotomy (where the weight distribution on the joint surfaces are altered; most common in the knee), or total joint replacement with a prosthesis. The latter is still most commonly used in the hip and knee, but the development of other artificial joints is proceeding rapidly.

Patient Handout 22-1

Degenerative Arthritis (Osteoarthritis)

Your doctor has determined that your joint pain is due to DEGENERATIVE ARTHRITIS (sometimes called DEGENERATIVE JOINT DISEASE or OSTEOARTHRITIS). You should be relieved to some extent: it means that you *don't* have a crippling disease like rheumatoid arthritis.

WHAT IS IT?

The surfaces of joints wear out. If you live long enough, all of your joints will wear out; so will everyone else's. But some people's joints wear out sooner than others, and doctors aren't sure why. They *do* know that joints that have been injured, or that have excessive strain placed on them, wear out much faster than others. (Examples would be an ankle that has been sprained severely, or the hips in someone that is overweight, or the base of the thumb in a mailman who pinches piles of letters all day.) In some people, the joints of the fingers begin wearing out for unknown reasons at a relatively young age. This wearing out of the joints is what is called DEGENERATIVE ARTHRITIS.

WHY DOES IT HURT?

Because the joint surface is rough instead of smooth, it hurts to move, at least after a while. A joint will feel stiff because the muscles around it tighten up.

WHAT CAN I DO ABOUT IT?

A lot of things!

1. Find a good balance of resting the joint and using it. The way to do that is to find by trial and error the amount of use that leads to the least pain during the day as a whole.

2. Aspirin not only helps take the pain away, but it actually reduces joint inflammation. If your joints only hurt occasionally, just take it when you need it; if they hurt a lot of the time, take a couple of aspirin four times a day on a regular basis. Never take it on an empty stomach. If it upsets your stomach anyway, try using a preparation of aspirin that has antacid mixed in. If it still leads to problems, or if you can't take aspirin because you are allergic to it, have a history of ulcers, or are on blood thinners, or if it doesn't seem to be doing the job, let your doctor know. There are other medications he or she can prescribe for you.

3. Heat is often helpful. Your doctor may give you some information on how to use it best.

4. If you are overweight and your problem is in your back, hips, knees, ankles, or feet, losing weight is one of the best things you can do for your joints.

5. If all these things don't give you sufficient relief, let your doctor know. He might prescribe splints, physical therapy, or recommend an injection. Rarely, surgery may be considered.

23

The Rheumatoid Diseases

Various patterns of joint involvement are more or less typical for certain diseases, but these patterns cannot be relied upon too heavily in arriving at a diagnosis. The evaluation of polyarthritis always requires a complete history and physical examination, since the cause is often a systemic illness, and findings in other organ systems can be crucial to determining the proper diagnosis. Laboratory tests are often also needed (see Chapter 2 for a discussion), and X rays (especially of the hands and spine) can be helpful. This chapter is by no means intended to be more than an outline of the more common arthritides; the reader is referred to a textbook of internal medicine or rheumatology for further information of these conditions.

Aside from the conditions reviewed in this chapter, GONOCOCCAL ARTHRITIS and VIRAL ARTHRITIS should be considered in acute oligoarthritis (Chapter 25); and SEPTIC ARTHRITIS (Chapter 25) or crystalline disease (Chapter 24) can sometimes affect more than one joint at a time. DEGENERATIVE ARTHRITIS and OSTEOARTHRITIS are covered in Chapter 22. ANKYLOSING SPONDYLITIS is discussed in Chapter 11, and JUVENILE RHEUMATOID ARTHRITIS and RHEUMATIC FEVER are covered in Chapter 19.

Rheumatoid Arthritis (*Common*)

Clinical Features

The usual onset is with involvement of several small or medium-sized joints, in a fairly symmetrical distribution.

The disease can, however, begin with a single inflamed joint.

There often are prodromal or associated systemic complaints such as stiffness and fatigue.

Morning stiffness is such a characteristic symptom that its duration can be used as an indicator of the level of disease activity.

The course of the disease usually is marked by spontaneous remissions and exacerbations. Severity of the disease varies tremendously, from patients with mild intermittent symptoms to those (fortunately less common) who are eventually crippled.

Some patients have rheumatoid nodules, which are subcutaneous and usually found at pressure points such as the olecranon.

Sjogren's syndrome (keratoconjunctivitis sicca, dry mouth and eyes) is seen in some patients; less common associated conditions include uveitis, episcleritis, vasculitis, pulmonary fibrosis, and polymyositis. *Felty's syndrome* is a rare combination of RA, splenomegaly, and neutropenia.

Females outnumber males three to one.

Once involved, a joint tends to remain involved as the disease spreads to other areas.

Examination early in the involvement of a joint will reveal warmth and tenderness; later on, synovial thickening will be palpable; and eventually loss of range of motion and various deformities will be evident.

The ESR is elevated and can be used to follow the level of disease activity.

The rheumatoid factor is positive in most patients with the disease but can sometimes be negative, especially early in the course.

A fair percentage of patients will have a positive ANA test.

Synovial fluid of involved joints will be typical for inflammatory conditions, with WBC counts in the 5,000 to 50,000 range and consisting mainly of PMN's.

X rays of involved joints show soft tissue swelling early, and then joint-space narrowing and progressive periarticular osteoporosis and bony erosions.

Pathophysiology

Although autoimmune factors no doubt play a part, the full picture of the etiology of this disease remains a mystery.

The synovium is inflamed and edematous. It becomes hypertrophic and spreads to cover the articular cartilage, destroying it and eventually the bone underlying it as well. This can then lead to adhesions in, and ankylosis of, the joint, as the supporting capsule is weakened by the inflammation.

Treatment

Local *heat* often gives symptomatic relief. For the hands, a paraffin bath is an easy way to provide this.

Range of motion and strengthening *exercises* are important in the maintenance of joint mobility and stability and thus the prevention of contractures and other deformities.

Referral to a physical therapist for training in and provision of the above modalities is indicated except in mildly affected individuals.

The first line of *analgesic/antiinflammatory medication* is aspirin. Usually the dose is pushed just to the point of toxicity and then cut back a bit to a maintenance level; the monitoring of blood salicylate levels is recommended. Buffered and time-release preparations or other salicylates can be used. See Chapter 3.

Where salicylates are not tolerated or do not give sufficient relief, other *nonsteroidal antiinflammatory drugs* can be tried. See Chapter 3.

Injection of steroid into severely affected joints can give significant relief for months. See the discussion in Chapter 3. Techniques are shown in the chapters on the individual joints.

In severe, progressive or unresponsive cases, *disease-suppressant medication* should be given. These are dangerous drugs, and the patient should be handled by a physician familiar with their use. Medicines now being used include gold salts, chloroquin, penicillamine, corticosteroids and various cytotoxic drugs.

Surgery, either synovectomy and/or procedures to correct deformities and contractures, is often necessary in severely affected joints.

Systemic Lupus Erythematosis (*Rare*)

This is a systemic autoimmune disease of unknown etiology. Fleeting joint pains as well as frank joint inflammation are the most common complaints. The arthritis rarely leads to any significant joint destruction or deformity. The disease also affects (in varying percentages of patients) serosal surfaces, the lymphoid system, the skin, the kidneys, and the CNS. Involvement of the latter two organ systems is often life-threatening. Corticosteroids are the first line of treatment. The reader is referred to texts of internal medicine or rheumatology for further information.

Psoriatic Arthritis (*Uncommon*)

Clinical Features

Most commonly there is an asymmetric polyarthritis involving a few small joints. The DIP joints are very commonly involved; nails adjacent to involved DIP joints are almost always pitted.

Although the activity of the joint disease tends to parallel activity of skin disease, either manifestation can appear first, so the absence of psoriasis of the skin does not rule out psoriatic arthritis.

About one-fourth of patients will have spondylitis, which is radiographically different from that seen in ankylosing spondylitis.

The sedimentation rate can be elevated; the rheumatoid factor is negative; uric acid is often elevated in psoriasis, so this should not lead to an erroneous diagnosis of gout. (Of course, sometimes two diseases are present.) The HLA-B27 Antigen is positive in 70% of cases in which spondylitis is a factor.

A destructive, progressive form, called *arthritis mutilans,* rarely occurs.

X ray changes are fairly characteristic, showing destruction of scattered small joints and occasionally the "pencil in cup" deformity at involved DIP's as well as fluffy periostitis.

Pathophysiology

The pathology of the synovial disease is similar to that of rheumatoid arthritis.

Treatment

Antiinflammatory medications and *physical modalities* are used as in rheumatoid arthritis.

In severe destructive disease, the patient should be referred for consideration of use of *cytotoxic agents* such as methotrexate.

Enteropathic Arthritis (*Rare*)

Clinical Features

About one-fifth of patients with ulcerative colitis and a smaller fraction of patients with regional enteritis have an associated arthritis, which takes one of two forms.

The more common type finds several large joints, more commonly of the lower extremity, becoming inflamed in succession. These attacks tend to resolve in weeks or months, leaving little or no residual damage. Activity of the joint disease tends to parallel that of the gastrointestinal disorder.

A smaller percentage of patients develop a spondylitis indistinguishable from that of ankylosing spondylitis.

Treatment

The peripheral arthritis is treated with *rest* and *antiinflammatory medication, and by treating the underlying GI disease.*

The spondylitis is managed the same as ankylosing spondylitis (covered in Chapter 11).

Reiter's Syndrome (*Uncommon*)

Clinical Features

The classical triad consists of arthritis, urethritis and conjunctivitis. This is a disease almost exclusively of young adult males.

The arthritis is usually asymmetric, and can involve a few or many joints, both large and small. Achilles tendonitis and plantar fascia inflammation often occur. Occasionally spondylitis is present. The arthritis usually persists for several weeks to several months and clears spontaneously. The spondylitis can be more persistent.

The urethritis must be distinguished by gram stain and culture from gonorrhea, which of course can also give an acute oligoarthritis.

The conjunctivitis is usually mild and transient; keratitis is rare.

Mucocutaneous lesions are common. *Keratoderma blennorrhagica* consists of painless pustules on the sole and palm, and is indistinguishable from pustular psoriasis. Occasionally painless shallow ulcerations of the buccal mucosa are found.

Recurrences are not uncommon.

Pathophysiology

It is likely that somehow the infecting organism of nonspecific urethritis (ureaplasma or chlamydia) leads to the other manifestations of the syndrome, but the mechanism by which this occurs remains unknown. The fact that the same syndrome sometimes occurs in association with shigella infections complicates the picture.

Treatment

Rest and *antiinflammatory* drugs should be used to suppress symptoms until the arthritis resolves.

Some physicians advocate a ten-day course of an antibiotic (tetracycline, erythromycin or trimethoprim/sulfa) even if the urethritis is no longer (or never was) present.

If any suspicion remains that the patient has gonococcal arthritis, treatment for this disease should be given as well. See Chapter 25.

Other Arthritides

An oligoarthritis can occasionally occur in *bacterial endocarditis*. Joint inflammation can sometimes precede the onset of the other features of *serum sickness*. Patients with *sarcoidosis* may have a migratory self-limited polyarthritis, or more likely a chronic arthritis related to granuloma deposition around the joint. *Behçet's syndrome* is a rare disease probably of viral origin, in which an intermittent oligoarthritis is a very common finding. (Other components of the syndrome are aphthous stomatitis, genital ulceration, iritis, and CNS involvement.) Attacks of *Familial Mediterranean Fever* can include synovitis of large joints. Various inborn errors of metabolism such as *Gaucher's disease* and *alcaptonuria* and inherited connective tissue disorders such as the *Ehlers-Danlos syndrome* can lead to joint pathology. Further discussion of these conditions is beyond the scope of this book.

24

The Crystalline Diseases

Gout (*Common*)

Clinical Features and Diagnosis

The condition is much more common in males.

The disease presents with one or more acute attacks. The first MTP joint is involved in fully one-half of first attacks (where it is called *podagra*); other commonly affected areas are the heel, instep, ankle and knee. The affected joint becomes suddenly red, warm, swollen and extremely painful. Untreated attacks last for days to weeks.

Attacks can be set off by fasting, alcohol binges, overuse, or the administration of diuretic or hypouricemic drugs; most often, however, no specific precipitant is noted.

The serum uric acid level is usually elevated, but the only way to make the diagnosis with absolute certainty is to tap the joint and find negatively birefringent uric acid crystals.

If hyperuricemia is untreated, over a decade or more the patient may develop *chronic gouty arthritis,* in which there are uric acid deposits in articular cartilage

348

and other soft tissue structures. *Tophi* also occur. These are urate deposits with surrounding inflammation, most often found in the pinna, tendons and superficial bursae.

A CBC and renal function tests should be performed.

About one-third of patients with gout will suffer at some time from urinary tract stones.

Pathophysiology

Uric acid is a normal by-product of nucleic acid metabolism. Hyperuricemia is due to uric acid overproduction or underexcretion, often of unknown cause. (The definition of abnormally high levels of uric acid in the blood must be an arbitrary one, since their distribution in the population follows a bell-shaped curve). Sometimes hyperuricemia is secondary to the use of thiazide diuretics or other drugs. Obesity and high-protein diets may be contributory. Infrequently the abnormality is a result of a known specific enzymatic defect. It can rarely be due to rapid purine turnover seen in the myeloproliferative disorders or chronic hemolytic anemia, or to decreased excretion in chronic renal disease.

In patients with hyperuricemia, urate exists as a supersaturated solution in most body fluids. Rapid fluctuations in level or possibly other factors then lead to the precipitation of crystals in a joint, i.e., an acute attack.

After many years, deposition of uric acid in various tissues can lead to chronic gouty arthritis and tophi.

Treatment

The acute attack should be treated with *elevation* and potent *antiinflammatory medication*. *Colchicine* has been the drug of choice for a long time, but in full therapeutic doses has more GI side effects than drugs such as indomethicin.

If hyperuricemia is thought to be drug-induced, consideration should be given to discontinuance of the suspected medication.

The higher the uric acid level, the more likely the patient will suffer recurrent attacks. With levels over 9 mg % they are almost certain to occur, and the development of chronic gouty arthritis in later years becomes a real possibility. With these facts in mind a decision must be made about whether to use *medication to lower acid levels*. If so, see Table 24-I. Levels should be followed periodically. Colchicine at a dose of 0.5 mg BID should be prescribed along with the hypouricemic drug for several months, to help prevent the acute attacks which can be precipitated by dropping uric acid levels.

Patients with *asymptomatic hyperuricemia* need not be treated unless levels are very high.

Table 24-I   Hypouricemic Drugs[a]

Drug	Adult dose	Comment
Allopurinol (XYLOPRIM, Burroughs Wellcome; LOPURIN, Boots; also available generically) 100 mg, 300 mg	100 to 600 mg/day (in divided doses if more than 300 mg/day)	Decreases uric acid production by inhibiting xanthine oxidase Decreases uric acid in urine, so useful in stone formers Not affected by salicylates Rare severe hypersensitivity reactions can occur: rash can be first sign
Probenecid (BENEMID, Merck, Sharp and Dohme; also available generically) 500 mg	250 mg BID for first week, then 500 mg BID (can push to 1 g BID if needed)	Salicylates interfere with action Increases uric acid excretion, so not to be used in stone formers; pushing fluids wise in all patients GI upset not uncommon Available in combination tablet with colchicine (COLBENEMID, MSD; or generic) for first few months of therapy
Sulfinpyrazone (ANTURANE, Ciba) 100 mg, 200 mg	100 mg BID to 200 mg BID (can push to 400 mg BID if needed)	Salicylates interfere with action GI upset not uncommon. Contraindicated with ulcer. Give with food or antacid Increases uric acid excretion, so same problems as probenecid

[a] See the package insert or Physician's Desk Reference for complete information before prescribing any drug.

Pseudogout (*Uncommon*)

Clinical Features and Diagnosis

There is acute inflammation of a single joint, most commonly the knee, in an older patient.

The episode resolves in a week or two. Rarely, more than one joint can be involved.

Examination of the synovial fluid reveals short positively birefringent crystals (calcium pyrophosphate).

This X ray will show linear calcification in articular and meniscal cartilage. (This is called *chondrocalcinosis*. While this is an invariable finding in pseudogout, it can also be seen in asymptomatic knees in older patients, and can be related to other conditions: hyperparathyroidism, gout, hemochromatosis, diabetes mellitus, Wilson's disease and acromegaly.)

Pathophysiology

Possibly the acute attack is precipitated by the release of some calcium crystals from cartilage into the joint space, where they initiate an inflammatory response.

Treatment

Elevation, limited-weight-bearing, and antiinflammatory medications are used.

25

Infection

JOINT INFECTION

Septic Arthritis (*Uncommon*)

(*Note:* Gonococcal arthritis is discussed separately below.)

Diagnosis

    The patient will present with an acutely inflamed joint, usually tender, red, and warm. But sometimes these signs are not dramatic, so all acutely and atraumatically inflamed joints must be tapped to rule out this possibility. Rarely, more than one joint will be involved.
    Fever and chills may or may not be present.

352

The peripheral WBC count will usually be elevated, but not always. Synovial fluid will show a high WBC count consisting mostly of PMN's. Gram stain may or may not reveal the causative organism. Cultures should be obtained for aerobic and anaerobic bacteria, for gonococcus, and if the picture is not clear, for fungi and tubercle bacilli as well.

Pathophysiology

Infection can involve the synovial lining or the joint cavity itself, or (usually) both. It may be secondary to hematogenous spread or to previous injection or aspiration of the joint; it may spread from nearby breaks in the skin. Occasionally it is an extension of a nearby osteomyelitis.

Treatment

Intravenous antibiotic therapy should be started immediately, specific for the organisms found by gram stain or culture. If no such clues are available at first, start with a semisynthetic penicillin active against staphylococcus, or a cephalosporin; in children add ampicillin to cover the strong possibility of hemophilus influenzae. (If ampicillin resistance is locally prevalent, add chloramphenicol as well until sensitivity results are in.) In infants, immune-suppressed patients, diabetics, and post-surgical cases, add gentamycin to the cephalosporin or semisynthetic pencillin (to cover gram-negative organisms). For dosages and recommended durations of treatment, see the PDR or internal medicine manuals or texts, or consult with a specialist in infectious disease.

Daily *needle aspiration* and *saline irrigation* of the joint probably speeds resolution. Determination of intraarticular antibiotic levels can be helpful. Intraarticular instillation of antibiotics is not necessary and can be very irritating.

Immobilization of the involved joint for a short period is symptomatically helpful. Physical therapy may be necessary to maintain or restore full range of motion.

If response to medical management is not prompt, with a significant decrease in pain and an afebrile condition being achieved in 24 to 48 hours, or if the hip or shoulder is involved *or,* if *any* joint is thought to be infected with staphylococcus, orthopedic consultation should be obtained.

Gonococcal Arthritis (*Common*)

Diagnosis

There is usually an asymmetric involvement of several large joints (or sometimes tendon sheaths).

Usually there is associated fever and a few small, pustular, hemorrhagic skin lesions.

A majority of patients with this syndrome of disseminated gonococcemia will have no symptoms of genital infection, so level of suspicion must remain high.

Urethral or cervical, rectal, and pharyngeal cultures should be obtained, as well as blood and synovial fluid cultures. A VDRL should also be done.

In the early stages of joint involvement, sometimes only the synovium is infected and no specific findings are present in the joint fluid.

Treatment

Currently recommended treatment regimens are as follows:*
Ampicillin/amoxicillin: ampicillin, 3.5 g, or amoxicillin, 3.0 g, orally, each with probenecid, 1.0 g, followed by ampicillin 0.5 g, or amoxicillin, 0.5 g, 4 times a day orally for 7 days.
Tetracycline: 0.5 g, orally 4 times a day for 7 days. (Tetracycline should not be used in pregnant women.)
Spectinomycin: 2.0 g, intramuscularly twice a day for 3 days. (This is the treatment of choice for disseminated infections caused by penicillin-resistant gonococcus.)
Erythromycin: 0.5 g, orally 4 times a day for 7 days.

Follow-up urethral or cervical, rectal, and pharyngeal cultures should be obtained after treatment. The patient should be instructed to inform his or her partners so that they may be treated, and a report should be sent to public health authorities.

Viral Arthritis (*Common*)

Diagnosis

An acute, usually mild oligo- or polyarthritis can accompany or sometimes precede the other symptoms of systemic viral infections. Most commonly this is seen with hepatitis A or B, with infectious mononucleosis, and with rubella (and sometimes after rubella immunization).

Treatment

The condition is self-limiting, and only symptomatic treatment need be prescribed.

*From the Center for Disease Control, *Morbidity and Mortality Weekly Report* **28**(2), Jan. 19, 1979.

Tuberculous, Mycotic, and Syphilitic Arthritis (*Rare*)

The reader is referred to textbooks of internal medicine or infectious disease for further discussion of these conditions, which usually present as a subacute or chronic monarthritis.

SOFT TISSUE INFECTION

Infected Superficial Bursae (*Common*)

Diagnosis

Usually the olecranon or patellar bursa is involved, with findings of tenderness, redness, and warmth.

The bursa should be tapped to obtain material for gram stain and culture and to decompress it.

Pathophysiology

Infection usually enters through nearby breaks in the skin. Staphylococcus aureus is by far the most common pathogen.

Treatment

An *antibiotic* effective against staph (semisynthetic penicillin or cephalosporin) should be prescribed, unless gram stain or culture reveals a different offending organism.

Hot soaks can be helpful.

The affected part should be immobilized.

If the course is prolonged, the patient must be observed for the spread of infection into nearby bone or joints.

Septic Tenosynovitis (*Rare*)

Diagnosis

Most commonly found in the hand or wrist, usually secondary to a puncture wound.

There is redness and extreme tenderness along the course of the tendon, and pain on its motion.

Treatment

Surgically opening the sheath is indicated, so *referral* to a hand surgeon should be made STAT.

Infections of the Various Spaces of the Hand (*Rare* except for infection in the volar tuft of the distal phalanx, called a *felon*)

These are potentially disastrous, and should be referred to a hand surgeon STAT. (PARONYCHIAS are excepted: see Figure 9-10.)

Cellulitis (*Common*)

Diagnosis

Diffuse inflammation with redness, warmth and tenderness is found over the involved area. BE CAREFUL: Never tap into a normal joint through superficial infection.
Systemic symptoms can occur.
A culture can be obtained by injecting sterile saline under the involved skin and then aspirating.

Pathophysiology

Infection is usually initiated through a break in the skin.
Streptococcus is the most common organism, especially if the border of the infected area is raised and purplish (called *erysipelas*). *Staphylococcus* is often present instead. Other organisms can be involved, especially in children (consider *hemophilus*) and in infants, diabetics and immune-suppressed individuals (consider *gram-negatives.*)

Treatment

If fever is absent, the involved area is small and the patient is an otherwise healthy adult, treatment can be with *elevation,* the application of *heat,* and *oral antibiotics* (use a cephalosporin or a semisynthetic penicillin effective against staphylococcus unless there is reason to suspect another organism or gram stain or culture results indicate another organism).

If resolution does not occur or the condition worsens, or if the patient has underlying disease, is an infant, has systemic symptoms or a large area is involved, admission for intravenous antibiotic therapy is advisable.

If loculated pus develops, it has to be drained.

Lymphangitis (*Common*)

Diagnosis

A red streak is noted going up an extremity, often but not always from a separately recognized site of infection.

Pathophysiology

This infection of the lymph duct is usually caused by hemolytic streptococcus.

Treatment

Guidelines are identical to those described under cellulitis.

BONE INFECTION

OSTEOMYELITIS should be suspected in any case of otherwise unexplained bony pain or tenderness; a bone scan will often be positive long before X ray changes are apparent. Since biopsy of bone is often required to obtain culture material, and surgical intervention is frequently needed in addition to medical treatment, this condition is best handled by the orthopedist.

References

GENERAL

American College of Surgeons Committee on Trauma: *Early Care of the Injured Patient*. ed 2. Philadelphia, Saunders, 1976.

Anderson JE: *Grant's Atlas of Anatomy*, ed 7. Baltimore, Williams & Wilkins, 1978.

Cailliet R: *Soft Tissue Pain and Disability*. Philadelphia, Davis, 1977.

DePalma AF: *The Management of Fractures and Dislocations: An Atlas*, 2 vols. Philadelphia, Saunders, 1970.

Cyriax J: *Textbook of Orthopedic Medicine*, 2 vols, ed 6. Baltimore, Williams & Wilkins, 1975.

Grant JCB: *Atlas of Anatomy*. ed 5. Baltimore, Williams & Wilkins, 1962.

Hollinshead WH:*Functional Anatomy of the Limbs and Back*, ed 4. Philadelphia, Saunders, 1976.

Iverson LD, Clawson DK: *Manual of Acute Orthopedic Therapeutics*. Boston, Little, Brown, 1977.

Kapandji IA: *The Physiology of the Joints*, 3 vols. Edinburgh, Churchill Livingstone, 1970.

LeVeau *et al: Biomechanics of Human Motion*, ed 2. Philadelphia, Saunders, 1977.

Ramamurti CP: *Orthopedics in Primary Care*. Baltimore, Williams & Wilkins, 1979.

Rockwood CA, Green D P: *Fractures*, 2 vols. Philadelphia, Lippincott, 1975.

Turek SL: *Orthopedics: Principles and their Application*, ed 3. Philadelphia, Lippincott, 1977.

CHAPTER 2

Daniels L, Worthingham C: *Muscle Testing: Techniques of Manual Examination*, ed 4. Philadelphia, Saunders, 1980.

Beetham WP *et al: Physical Examination of the Joints*. Philadelphia, Saunders, 1965.

Edeiken J, Hodes PJ: *Roentgen Diagnosis of Diseases of the Bone*, 2 vols, ed 2. Baltimore, Williams & Wilkins, 1973.

Forrester DM, Brown JC, Nesson JW: *The Radiology of Joint Disease*, Philadelphia, Saunders, 1978.

Licht S (ed): *Electrodiagnosis and Electromyography,* ed 3. New Haven, Elizabeth Licht, 1971.
Murray RO, Jacobsen HG: *The Radiology of Skeletal Disorders,* 4 vols, ed 2. Edinburgh, Churchill Livingstone, 1977.
Watanabe M, Takeda S, Ikeuchi H: *Atlas of Arthroscopy,* ed 3. Tokyo, Igaku-Shoin, 1979.

CHAPTER 3

Baker CE: *Physicians Desk Reference.* ed 34. Oradell, New Jersey, Medical Economics Co, 1980.
Baker CE: *Physicians Desk Reference for Nonprescription Drugs,* ed 1. Oradell New Jersey, Medical Economics Co, 1980.
Rusk HA: *Rehabilitation Medicine,* ed 4. St. Louis, Mosby, 1977.
Steinbrocker O, Neustadt DH: *Aspiration and Injection Therapy in Arthritis and Musculoskeletal Disorders,* Hagerstown, Maryland, Harper, 1972.

CHAPTER 4

Cailliet R: *Neck and Arm Pain.* Philadelphia, Davis, 1964.
Bateman JE: *The Shoulder and Neck,* ed 2. Philadelphia, Saunders, 1978.
Jackson R: *The Cervical Syndrome,* ed 3. Springfield, Illinois, Charles C Thomas, 1971.

CHAPTER 5

Cailliet R: *Neck and Arm Pain.* Philadelphia, Davis, 1964.
Chusid JG: *Correlative Neuroanatomy and Functional Neurology,* ed 15. Los Altos, California, Lange, 1973.
Hale MS: *A Practical Approach to Arm Pain,* Springfield, Illinois, Charles C Thomas, 1971.
Kopell HP, Thompson WAL: *Peripheral Entrapment Neuropathies.* Huntington, Robert E. Krieger, New York, 1976.

CHAPTER 6

Bateman JE: *The Soulder and Neck,* ed 2. Philadelphia, Saunders, 1978.
Cailliet R: *Shoulder Pain.* Philadelphia, Davis, 1966.

CHAPTER 7

Lister GD, Belsole RB, Keinert HE: The radial tunnel syndrome. *J Hand Surg* 4(1):52, 1979.

CHAPTER 8

Cailliet R: *Hand Pain and Impairment.* Philadelphia, Davis, 1975.

CHAPTER 9

Cailliet R: *Hand Pain and Impairment.* Philadelphia, Davis, 1975.
Ramamurti CP: *Orthopedics in Primary Care.* Baltimore, Williams & Wilkins, 1979.
Ruby LK: *Common Hand Injuries in the Athlete,* Symposium on Sports Injuries, Orthopedic Clinics of North America, 11, 4. Philadelphia, Saunders, 1980.

CHAPTER 11

Cailliet R: *Low Back Pain Syndrome,* ed 2. Philadelphia, Davis, 1968.
DePalma AF, Rothman RH: *The Intervertebral Disc.* Philadelphia, Saunders, 1970.
Drugs for postmenopausal osteoporosis. *Medical Letter on Drugs and Therapeutics* 22(11), 1980.
Helfet AJ, Gruebel Lee DM: *Disorders of the Lumbar Spine.* Philadelphia, Lippincott, 1978.
MacNab I: *Backache.* Baltimore, Williams & Wilkins, 1977.

CHAPTER 13

Chusid JG: *Correlative Neuroanatomy and Functional Neurology,* ed 15. Los Altos, California, Lange, 1973.
Cailliet R: *Foot and Ankle Pain.* Philadelphia, Davis, 1968.
Kopell HP, Thompson WAL: *Peripheral Entrapment Neuropathies.* Huntington, New York, Robert E. Krieger, 1976.

CHAPTER 14

Strange FGSC: *The Hip.* Baltimore, Williams & Wilkins, 1965.

CHAPTER 15

Cailliet R: *Knee Pain and Disability.* Philadelphia, Davis, 1973.
Helfet AJ: *Disorders of the Knee.* Philadelphia, Lippincott, 1974.
Smillie IS: *Injuries of the Knee Joint,* ed 5. Edinburgh, Churchill Livingstone, 1978.

CHAPTER 17

Cailliet R: *Foot and Ankle Pain.* Philadelphia, Davis, 1968.
Stover CN: Air stirrup management of ankle injuries in the athlete, *Am J Sportsmed* 8:5, 1980.

CHAPTER 18

Cailliet R: *Foot and Ankle Pain.* Philadelphia, Davis, 1968.
Ramamurti CP: *Orthopedics in Primary Care.* Baltimore, Williams & Wilkins, 1979.

CHAPTER 19

American Academy of Pediatrics: *School Health: A Guide for Health Professionals.* 1977.
Green M, Haggerty RJ: *Ambulatory Pediatrics II.* Philadelphia, Saunders, 1977.
Hoekelman R *et al: Principles of Pediatrics: Health Care of the Young.* New York, McGraw-Hill, 1978.
Lovell WW, Winter RB (eds): *Pediatric Orthopedics,* 2 vols. Philadelphia, Lippincott, 1978.
Symposium of common orthopedic problems, *Pediatric Clinics of North America.* Philadelphia, Saunders, 1977.
Vaughan VC, McKay RJ: *Nelson's Textbook of Pediatrics.* ed 10. Philadelphia, Saunders, 1975.

CHAPTER 20

Anderson B: *Stretching.* Bolinas, California, Shelter Publications, 1980.
Brody DM: *Running Injuries.* Ciba Clinical Symposia, Vol 32(4), 1980.
Clancy WG (ed): Symposium on runner's injuries, *Am J Sportsmed* 8:2, 4, 1980.
Daniel DM: *Lower Limb Maladies in the Runner.* A Course on Orthopedics for the Primary Physician, UCSD, San Diego, March 1980.
James SL, Bates BT, Osternig LR: Injuries to runners, *Am J Sportsmed* 8:5, 1980.
Mann RA, Hagy J: Biomechanics of walking, running and sprinting, *Am J Sportsmed* 8(5), 1980.
Pagliano J, Jackson D: The ultimate study of running injuries, *Runner's World* 15:11, 1980.

CHAPTER 21

Adams RD: *Diseases of Muscle,* ed 3. Hagerstown, Maryland, Harper, 1975.

CHAPTER 22

Arthritis Foundation. Primer on the Rheumatic Diseases, ed 7. New York, 1973.
Katz WA: *Rheumatic Diseases: Diagnosis and Management.* Philadelphia, Lippincott, 1977.
Kelly WN *et al: Textbook of Rheumatology.* Philadelphia, Saunders, 1981.
McCarty DJ: *Arthritis and Allied Conditions: A Textbook of Rheumatology,* ed 9. Philadelphia, Lea & Febiger, 1979.
Wintrobe MM *et al: Harrisons Principles of Internal Medicine,* ed 7. New York, McGraw-Hill, 1974.

CHAPTER 23

Arthritis Foundation, Primer on the Rheumatic Diseases, ed 7. New York, 1973.
Katz WA: *Rheumatic Diseases: Diagnosis and Management.* Philadelphia, Lippincott, 1977.
Kelly WN *et al: Textbook of Rheumatology.* Philadelphia, Saunders, 1981.
McCarty DJ: *Arthritis and Allied Conditions: A Textbook of Rheumatology,* ed 9. Philadelphia, Lea & Febiger, 1979.
Wintrobe MM *et al: Harrisons Principles of Internal Medicine,* ed 7. New York, McGraw-Hill, 1974.

CHAPTER 24

Arthritis Foundation. Primer on the Rheumatic Diseases, ed 7. New York, 1973.

Baker CE: *Physicians Desk Reference,* ed 3,4. Oradell, New Jersey, Medical Economics Co, 1980.

Katz WA: *Rheumatic Diseases: Diagnosis and Management.* Philadelphia, Lippincott, 1977.

Kelly WN *et al: Textbook of Rheumatology.* Philadelphia, Saunders, 1981.

McCarty DJ: *Arthritis and Allied Conditions: A Textbook of Rheumatology* ed 9. Philadelphia, Lea and Febiger, 1979.

Talbott JH, Yu T-F: *Gout and Uric Acid Metabolism.* New York, Stratton, 1976.

Wintrobe MM *et al: Harrisons Principles of Internal Medicine,* ed 7. New York, McGraw-Hill, 1974.

CHAPTER 25

Gardner P, Provine HT: *Manual of Acute Bacterial Infections.* Boston, Little, Brown, 1975.

Iverson Ld, Clawson DK: *Manual of Acute Orthopedic Therapeutics.* Boston, Little, Brown, 1977.

Index

A

Abnormal curvature of back, 282
  evaluation, 280–281
  structural scoliosis, 281
  treatment, 281
Accessory bones, foot, 252
Achilles tendonitis, 312
  diagnosis, 239, 302
  pathophysiology, 239, 303
  treatment, 239, 303
Acromioclavicular joint arthritis, 69
  diagnosis, 75
  pathophysiology, 76
  treatment, 76
Acromioclavicular joint sprain
  diagnosis, 78
  pathophysiology, 78
  treatment, 78
Acute foot strain
  diagnosis, 232
  treatment, 232
Acute low back strain, 279
  diagnosis, 130
  pathophysiology, 130–131
  treatment, 131
Acute lumbosacral radiculitis, 161, 279
  diagnosis, 131–132
  pathophysiology, 132
  treatment, 132
Acute patellar dislocation
  diagnosis, 201
  treatment, 201
Acute torticollis, 32, 283
  diagnosis, 34
  pathophysiology, 34
  treatment, 34

Adductor strain, 321
  diagnosis, 320
  pathophysiology, 320
  treatment, 320
Adhesive capsulitis, 69, 73
  diagnosis, 76
  pathophysiology, 77
  treatment, 77
Amphiarthrodal joint, definition, 2
Ankle
  anatomy, 210–211
  how to tape, 228
  instability
    evaluation, 222
    laxity of lateral ligaments, 222–223
  lumps, 223
  pain, 222
    anterior tibialis and extensor tendonitis, 215
    degenerative arthritis of ankle, 216–217
    degenerative arthritis of subtalar joint, 215–216
    examination, 212–213
    history, 211–212
    peroneal tendonitis, 214–215
    posterior tibialis tendonitis, 212
  sites of tenderness in, 213
  sprains, 303
    examination, 217–218
    first degree, 220
    second degree, 220–221
    third degree, 221–222
    X ray, 218–220
Ankylosing spondylitis, 122, 129, 342
  diagnosis, 136–137
  treatment, 137
Anserine tendonitis
  diagnosis, 195, 317